# Artificial Intelligence Paradigms for Application Practice

This book proposes practical application paradigms for deep neural networks, aiming to establish best practices for real-world implementation.

Over the past decade, deep neural networks have made significant progress. However, effectively applying these networks to solve various practical problems remains challenging, which has limited the widespread application of artificial intelligence. *Artificial Intelligence Paradigms for Application Practice* is the first book to comprehensively address implementation paradigms for deep neural networks in practice. The authors begin by reviewing the development of artificial neural networks and provide a systematic introduction to the tasks, principles, and architectures of deep neural networks. They identify the practical limitations of deep neural networks and propose guidelines and strategies for successful implementation. The book then examines 14 representative applications in urban planning, industrial production, and transportation. For each case, the authors present a landing paradigm that effectively addresses practical challenges supported by illustrations, background information, related work, methods, experiments, and conclusions. The experimental results validate the effectiveness of the proposed implementation approaches.

The book will benefit researchers, engineers, undergraduate, and graduate students interested in artificial intelligence, deep neural networks, large models, stable diffusion models, video surveillance, smart cities, intelligent manufacturing, intelligent transportation, and other related areas.

**Shiguo Lian** earned his PhD from Nanjing University of Science and Technology, China. He currently serves as Chief Scientist at the Data Science & Artificial Intelligence Research Institute of China Unicom and Chief Engineer of AI Technology at China Unicom. He is a member of the IEEE Multimedia Communications and Computational Intelligence Technical Committees. His research focuses on visual recognition, multimodal large models, robotics, and multimodal interactions.

**Zhaoxiang Liu** earned his PhD from the College of Information and Electrical Engineering at China Agricultural University, China. He currently serves as Director of AI Research at the Data Science & Artificial Intelligence Research Institute, China Unicom. His research interests include artificial intelligence, large language models, multimodal large models, deep learning, computer vision, and embodied AI.

# Artificial Intelligence Paradigms for Application Practice

Shiguo Lian and Zhaoxiang Liu

CRC Press
Taylor & Francis Group
Boca Raton London New York

CRC Press is an imprint of the
Taylor & Francis Group, an **informa** business

Designed cover image: Shutterstock (2520834961)

First edition published 2026
by CRC Press
2385 NW Executive Center Drive, Suite 320, Boca Raton FL 33431

and by CRC Press
4 Park Square, Milton Park, Abingdon, Oxon, OX14 4RN

*CRC Press is an imprint of Taylor & Francis Group, LLC*

© 2026 Shiguo Lian and Zhaoxiang Liu

ISBN: 978-1-041-08226-2 (hbk)
ISBN: 978-1-041-08348-1 (pbk)
ISBN: 978-1-003-64497-2 (ebk)

DOI: 10.1201/9781003644972

Typeset in Nimbus font
by KnowledgeWorks Global Ltd.

# Contents

# Acknowledgments

This book would not have been possible without the support and contributions of numerous individuals. We would like to express our deepest gratitude to our colleagues at Data Science & Artificial Intelligence Research Institute for their invaluable assistance throughout this project. They are: Kaikai Zhao, Xiang Wang, Huan Hu, Zezhou Chen, Zipeng Wang, Wenjing Zhang, Xiang Liu, Minjie Hua, Chao Tan, and Kai Wang. We really appreciate their help in collecting background and supplementary materials, providing precious source data, case studies, and revision advice, editing visual materials, refining references and citations, polishing language, and proofreading. Their efforts, though often behind the scenes, were essential to the completion of this work.

Finally, we would like to acknowledge the broader community of researchers and practitioners in the fields of AI and computer vision, whose groundbreaking work has inspired and informed this book. This project stands on the shoulders of their collective efforts.

# CHAPTER 1

# Introduction to Artificial Intelligence for Practical Application

## 1.1 THE DEVELOPMENT HISTORY OF ARTIFICIAL NEURAL NETWORKS

The artificial neural network (ANN) is a mathematical model that mimics the structure of synaptic connections in the brain for information processing. Its emergence was inspired by the organization of biological neurons, and its development has undergone significant fluctuations over time. Figure 1.1 illustrates the development history of artificial neural networks.

The development of artificial neural networks has progressed through three significant waves. The first wave began in 1943 when psychologist Warren McCulloch and logician Walter Pitts introduced the concept of artificial neural networks and developed a mathematical model of an artificial neuron [150], initiating an era of neural network research. A notable contribution during this period was from American neuroscientist Frank Rosenblatt, who proposed the concept of "perceptrons," machines capable of simulating human perception [188]. However, in 1969, Marvin Minsky, often called "the father of artificial intelligence," and Seymour Papert highlighted the limitations of single-layer perceptrons, particularly their inability to solve problems like the XOR problem [153]. This critique led to a decline in neural network research.

The second wave of artificial neural networks emerged in the 1980s as many previously encountered problems were gradually addressed. A significant breakthrough occurred in 1986 when David Rumelhart, Geoffrey Hinton, and Ronald Williams applied the backpropagation (BP) algorithm to neural network training, proposing an effective method for learning the weights of hidden layers in multi-layer neural networks [192]. They also introduced the Sigmoid function, which added nonlinearity to neural networks. This period marked a resurgence in neural network research. Despite these advances, second-generation neural networks still faced many limitations. With the advent of the Support Vector Machine (SVM) algorithm [37], interest in neural networks waned once again.

DOI: 10.1201/9781003644972-1

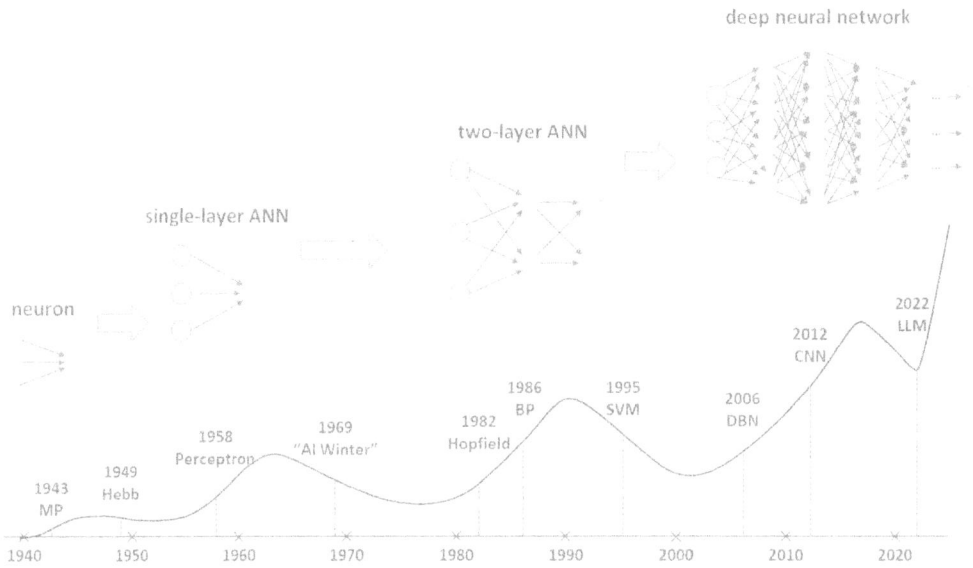

Figure 1.1   The development history of artificial neural networks.

The third wave of artificial neural networks began in 2006 with the introduction of the Deep Belief Network (DBN) by Geoffrey Hinton and his colleagues [80]. This deep network model emphasized the depth of the model structure and the importance of feature learning. Through layer-by-layer feature transformation, deep neural networks could convert the original features of the sample into a new feature space, making classification or prediction more accurate and efficient. The success of deep models also depended on the availability of large datasets and advanced parallel processing hardware.

During this period, several classic networks emerged. In 2012, Alex Krizhevsky et al. developed AlexNet, a groundbreaking model for image classification [112]. In 2014, Ian Goodfellow and his colleagues proposed the Generative Adversarial Network (GAN) [70]. In 2015, Olaf Ronneberger et al. created the U-Net for biomedical image segmentation [187], and Kaiming He et al. introduced the Residual Neural Network (ResNet) [78]. In 2017, the Google Brain team introduced the Transformer model, which revolutionized natural language processing (NLP) [219]. In 2018, Jacob Devlin et al. at Google published BERT (Bidirectional Encoder Representations from Transformers) [44]. In 2020, OpenAI released the Generative Pre-trained Transformer 3 (GPT-3) [16]. In 2022, the CompVis group at Ludwig Maximilian University Munich proposed the Stable Diffusion model [186], and OpenAI developed and released ChatGPT [159].

These advancements have markedly propelled the field of artificial neural networks, demonstrating their potential in various applications and cementing their pivotal role in contemporary artificial intelligence research. In the following sections, we will delve into the details of two paramount neural network architectures: CNN and Transformer.

## 1.2  CONVOLUTIONAL NEURAL NETWORKS

Convolutional Neural Networks (CNNs), inspired by the organization of the visual cortex in animals, were first proposed by Yann LeCun and his collaborators in the late 1980s and early 1990s [118, 119]. A landmark achievement in this field was the development of LeNet-5 in 1998, a CNN specifically designed for recognizing handwritten digits [119]. LeNet-5 demonstrated exceptional performance on the MNIST dataset, setting a foundation for significant advancements in CNN research and applications.

The convolutional layer is the foundational component of CNNs. It employs a set of learnable filters, or convolution kernels, to scan input data and extract local features. The convolution kernel slides across the input data, performing element-wise multiplication and summation at each position to compute a new value. For instance, when using a $3 \times 3$ convolution kernel on a $5 \times 5$ region of an image, the kernel slides over the image and performs a dot product operation with each corresponding $3 \times 3$ image block, producing new values that represent local features (see Figure 1.2).

In the case of color images, which have three channels (red, green, and blue, or RGB), the convolution kernel is also designed with three corresponding channels. During the convolution process, the kernel operates on each image channel separately, computing the results for each and summing them to generate the value for the output channel. By employing multiple convolution kernels, a variety of features can be extracted, increasing the number of output channels.

To formalize this, consider an input feature map with dimensions $(1, i_C, i_H, i_W)$, where $i_C$ is the number of input channels, and $i_H$ and $i_W$ are the height and width of the input. A convolution kernel has dimensions $(1, i_C, k, k)$, where $k$ represents the kernel size. When a single kernel is applied, it produces an output feature map with

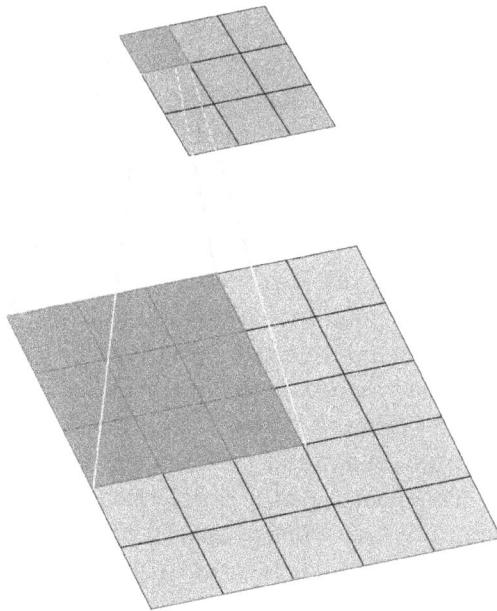

Figure 1.2  Illustration of the convolution operation.

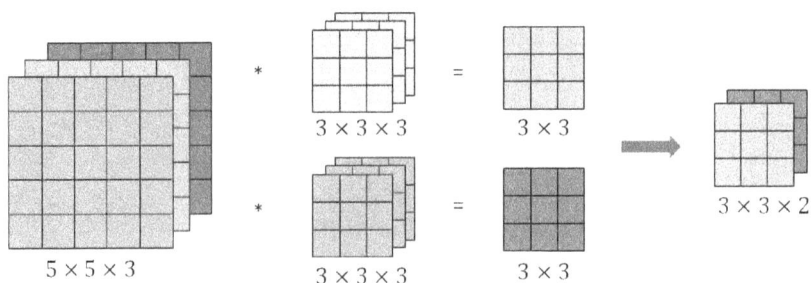

**Figure 1.3**  Convolution operation on a $5 \times 5 \times 3$ image using two $3 \times 3 \times 3$ kernels, producing a $3 \times 3 \times 2$ feature map with spatial resolution $3 \times 3$ and two output channels.

dimensions $(1, 1, o_H, o_W)$, where $o_H$ and $o_W$ are the height and width of the output. If the convolutional layer contains $o_C$ kernels, the resulting output feature map will have dimensions $(1, o_C, o_H, o_W)$, capturing multiple feature representations across the $o_C$ channels. Figure 1.3 shows an example.

CNNs revolutionized computer vision by introducing the concept of convolutional layers, which enable automatic feature extraction from images. This feature extraction capability drastically reduced the need for manual feature engineering, allowing neural networks to learn hierarchical representations directly from raw pixel data.

The architecture of a typical CNN consists of multiple layers, including convolutional layers, pooling layers, and fully connected layers. Convolutional layers apply a filter (or kernel) across the input data to produce feature maps, extracting spatial features such as edges, textures, and shapes. Following the convolution operation, an activation function, commonly the Rectified Linear Unit (ReLU), is applied to introduce non-linearity into the model. Pooling layers then reduce the spatial dimensions of the feature maps, thereby decreasing the number of parameters and computations in the network. Common pooling operations include max pooling and average pooling, both of which provide spatial invariance, enabling the network to recognize objects in images regardless of their position. After several convolutional and pooling layers, high-level reasoning is performed by fully connected layers, which take the output of the convolutional layers and produce the final classification or regression output. Additional techniques commonly used in neural networks include dropout and batch normalization. Dropout is a regularization method that prevents overfitting by randomly setting a fraction of the input units to zero during each training update. Batch normalization normalizes the inputs of each layer, which stabilizes and accelerates the training process. The architecture of LeNet-5 [119] is illustrated in Figure 1.4.

CNNs have demonstrated remarkable performance across various computer vision tasks, including image classification, object detection, semantic segmentation, and image generation. Their ability to capture spatial hierarchies of features has led to significant advancements in fields such as medical imaging, autonomous vehicles, and industrial quality inspection.

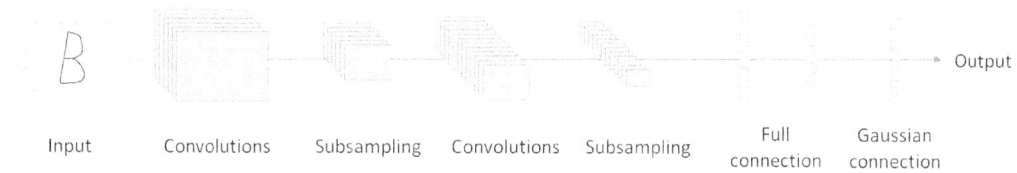

**Figure 1.4**  Architecture of LeNet-5. (The figure is inspired by and drawn based on [119].)

## 1.3  TRANSFORMER

The Transformer model was introduced by Vaswani et al. in the seminal paper "Attention is All You Need," presented at the Advances in Neural Information Processing Systems (NeurIPS) conference in 2017 [219]. This model is designed for sequence transduction tasks and relies exclusively on attention mechanisms, completely eliminating the need for recurrence and convolutions. A primary attribute of the Transformer model is its capacity to capture long-range dependencies in sequential data while facilitating parallel processing. The introduction of Transformers has profoundly impacted the field of NLP, with their applications now extending beyond text to encompass domains such as speech and image processing.

Transformer networks are built on several key concepts and components (see Figure 1.5). The self-attention mechanism is central, allowing the model to weigh the importance of different words in a sentence relative to each other, capturing dependencies regardless of their distance in the sequence. Scaled dot-product attention computes scores for each word relative to every other word, normalizes these scores, and uses them to produce weighted sums of the input representations. Unlike Recurrent Neural Networks (RNNs) or CNNs, Transformers do not inherently account for the order of input sequences; thus, positional encoding is added to input embeddings to provide information about the position of each word. The encoder-decoder architecture comprises multiple identical layers. The encoder, with multi-head self-attention mechanisms and fully connected feed-forward networks, processes the input sequence and generates continuous representations. The decoder, with an additional attention mechanism over the encoder's output, focuses on relevant parts of the input sequence when generating the output. Multi-head attention captures different aspects of the input data by having each head perform its own self-attention operation, then concatenating and linearly transforming their outputs. Each sub-layer in the Transformer is followed by layer normalization and residual connections, which facilitate gradient flow and improve training stability.

Transformer-based neural networks have become fundamental to the field of artificial intelligence, particularly in NLP, due to their ability to manage sequential data with remarkable efficiency and accuracy. Several key Transformer-based models have made significant contributions to AI research and applications, including BERT (Bidirectional Encoder Representations from Transformers) [44], GPT (Generative

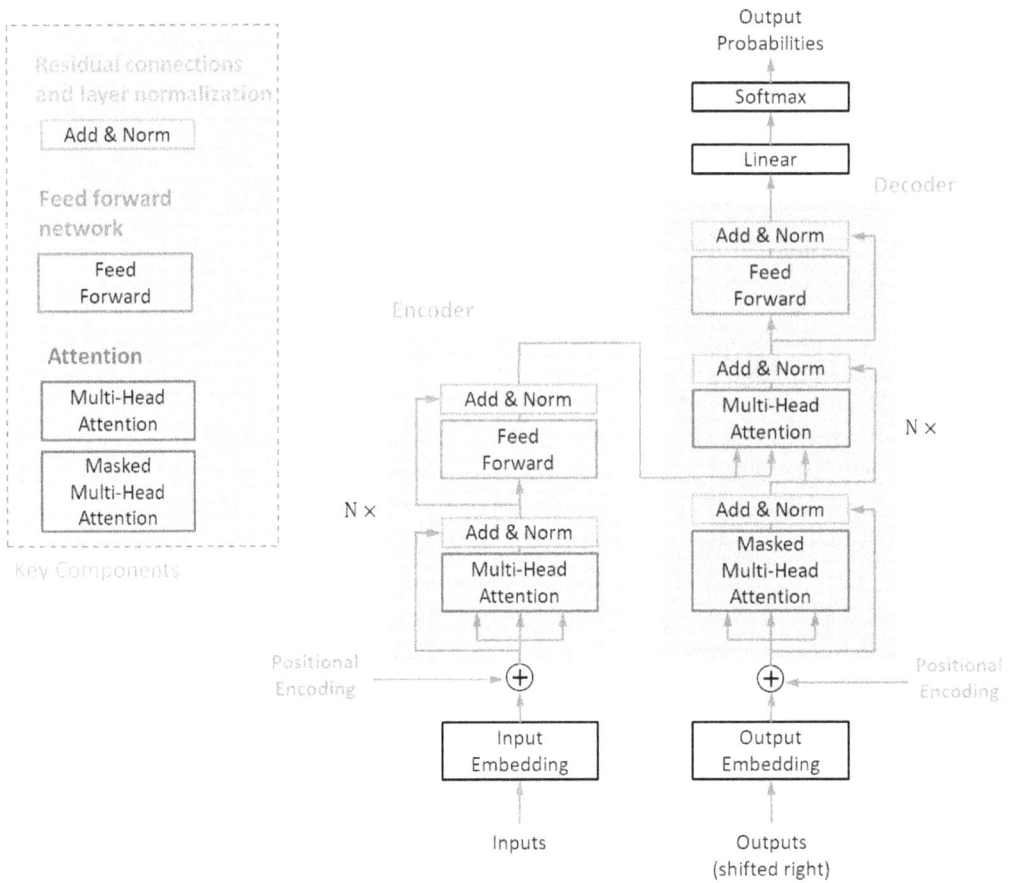

**Figure 1.5** The key components of the Transformer model: attention, feed forward network, residual connections, and layer normalization. (The figure is inspired by and drawn based on [219].)

Pre-trained Transformer) [174], Transformer-XL [40], and T5 (Text-to-Text Transfer Transformer) [176], among others.

Inspired by the success of Transformers in NLP, researchers began exploring their potential in vision tasks. A significant breakthrough occurred with the introduction of the Vision Transformer (ViT) by Dosovitskiy et al. in 2020 [48]. ViT applies the Transformer architecture to image classification by treating image patches as sequences of tokens, analogous to words in a sentence. ViT demonstrated that Transformers could achieve state-of-the-art performance on image classification benchmarks when pretrained on large datasets. Following ViT, several variants and enhancements, such as Data-efficient Image Transformers (DeiT) [216], Swin Transformer [142], and Pyramid Vision Transformer (PVT) [231], were proposed to improve performance and efficiency.

Traditionally, CNNs have dominated computer vision tasks due to their ability to capture spatial hierarchies through convolutional layers. However, CNNs have limitations, particularly in modeling long-range dependencies and processing efficiency.

While Transformers have rapidly gained traction in the vision field, they also present several challenges, such as high resource demand, significant memory consumption, the need for large datasets, and time-consuming training processes. Additionally, Transformers lack inherent spatial hierarchies. Ongoing research aims to address these challenges by developing more efficient and effective Transformer variants, hybrid models that combine the strengths of CNNs and Transformers, and advanced training techniques that reduce the data and computational requirements of these models.

## 1.4 TYPICAL APPLICATIONS OF NEURAL NETWORKS IN VISION

Neural networks have revolutionized image and video analysis across numerous domains, showcasing their versatility in computer vision tasks. The subsequent discussion will explore several typical applications and networks in this field.

### 1.4.1 Image Classification

Neural networks can classify images into predefined categories, facilitating tasks such as object recognition, facial recognition, and content-based image retrieval. Below are two prominent networks that have been instrumental in driving these advancements.

#### 1.4.1.1 ResNet (Residual Network)

ResNet is a highly influential CNN architecture proposed by Kaiming He, Xiangyu Zhang, Shaoqing Ren, and Jian Sun in their 2015 paper "Deep Residual Learning for Image Recognition." [78] The key contribution of ResNet is its effective solution to the problem of training extremely deep neural networks by addressing the "vanishing gradient problem." This challenge often impedes the training of deep networks, making it difficult for gradients to propagate through many layers. ResNet introduced a novel approach using "skip connection" or "residual connection" (see Figure 1.6), which allows gradients to flow more easily through the network.

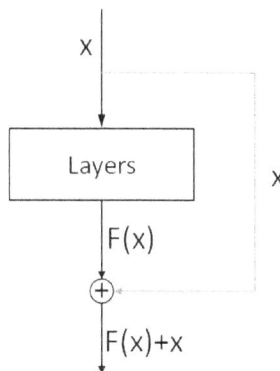

Figure 1.6   Illustration of the "skip connection" or "residual connection." (The figure is inspired by and drawn based on [78].)

ResNet features an innovative architecture characterized by the use of residual blocks. Each residual block consists of a series of convolutional layers, typically followed by batch normalization and ReLU activation. What distinguishes ResNet is the identity shortcut connection that skips one or more layers, directly adding the input of the block to its output. This shortcut connection allows gradients to flow through the network more effectively, making it feasible to train much deeper networks than previously possible.

ResNet's architecture and the concept of residual learning have fundamentally changed the design of deep neural networks, establishing it as a cornerstone in modern deep learning research and applications. ResNet had a profound impact on image classification and other computer vision tasks by demonstrating that extremely deep networks, with hundreds or even thousands of layers, could be trained effectively. The residual learning approach has been widely adopted and adapted in various neural network architectures beyond image classification, including object detection, semantic segmentation, and NLP.

### 1.4.1.2  ViT (Vision Transformer)

The ViT was introduced by Alexey Dosovitskiy and his colleagues in their seminal 2020 paper, "An Image is Worth 16x16 Words: Transformers for Image Recognition at Scale." [48] Drawing inspiration from the success of Transformer models in NLP, ViT adapts the Transformer architecture to process image data, offering a novel methodology for image classification.

ViT's architecture diverges from traditional CNNs by eliminating convolutions entirely. Instead, it employs the Transformer model, originally designed for sequential data such as text, for image classification tasks. ViT begins by dividing an image into a grid of non-overlapping patches, typically 16x16 pixels in size. These patches are then flattened into vectors and linearly embedded, incorporating positional embeddings to retain information about the spatial relationships between patches. This sequence of embedded patches is fed into a standard Transformer encoder, which comprises multiple layers of multi-head self-attention and feed-forward neural networks (see Figure 1.7).

The introduction of ViT illustrated that Transformers could attain cutting-edge performance on image classification benchmarks, rivaling the established dominance of CNNs. ViT achieved exceptional results when pre-trained on extensive datasets like ImageNet-21k or JFT-300M and subsequently fine-tuned on smaller datasets, demonstrating its effectiveness in managing high-dimensional visual data. This represented a significant paradigm shift in image classification methodologies, underscoring the potential of attention mechanisms in visual tasks.

By demonstrating that Transformers can outperform traditional CNNs in image classification tasks, ViT has prompted researchers to investigate non-convolutional architectures for a variety of computer vision applications. ViT effectively bridges the gap between NLP and computer vision, proposing a unified framework for processing different data types with a single model architecture. The success of ViT is particularly evident when scaled with large training datasets and significant computational

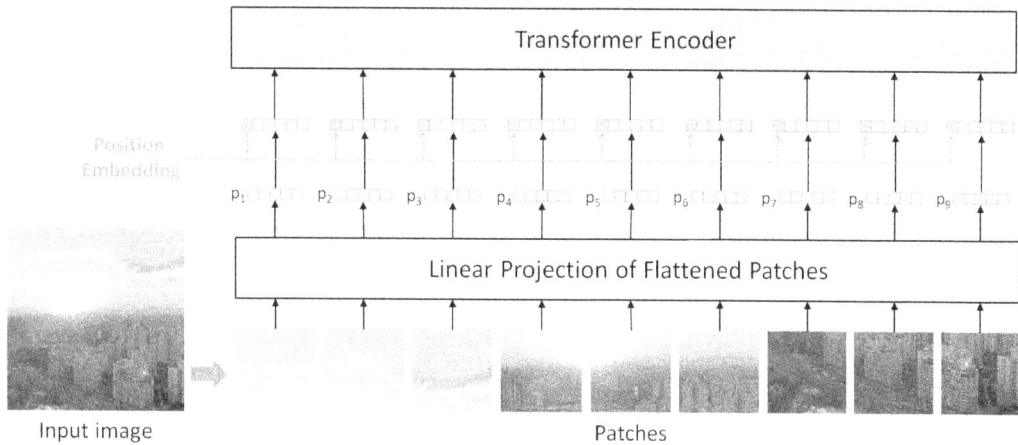

**Figure 1.7** The ViT model receive image as input. (The figure is inspired by and drawn based on [48].)

resources, highlighting the importance of dataset size and model capacity in achieving high performance in deep learning. ViT's success has also spurred the development of numerous variants and improvements, such as Data-efficient Image Transformers (DeiT) [216], which reduce the dependency on large datasets, and hybrid models that integrate the strengths of CNNs and Transformers [107].

## 1.4.2 Object Detection

Neural networks can detect and localize objects within images, providing bounding boxes around identified objects along with their corresponding class labels. This ability is critical for numerous real-world tasks, such as autonomous driving, surveillance, healthcare, and robotics. In the deep learning era, the development of object detection has primarily focused on two main approaches: two-stage algorithms and one-stage algorithms.

The key distinction between these approaches is that two-stage algorithms first generate proposals (pre-selected regions that may contain objects) and then perform fine-grained object detection within these proposals. In contrast, one-stage algorithms directly predict object classifications and locations from the features extracted by the network, bypassing the proposal generation step. Below, we introduce a representative network for each approach.

### 1.4.2.1 Faster R-CNN (Faster Region-Based Convolutional Neural Network)

Faster R-CNN was proposed by Shaoqing Ren, Kaiming He, Ross Girshick, and Jian Sun in their 2015 paper "Faster R-CNN: Towards Real-Time Object Detection with Region Proposal Networks" [182]. It represents a significant evolution from its predecessors, R-CNN [64], and Fast R-CNN [63]. R-CNN pioneered the concept of region proposals, employing selective search to generate a multitude of region proposals, which were then subjected to feature extraction and classification. Fast R-CNN built

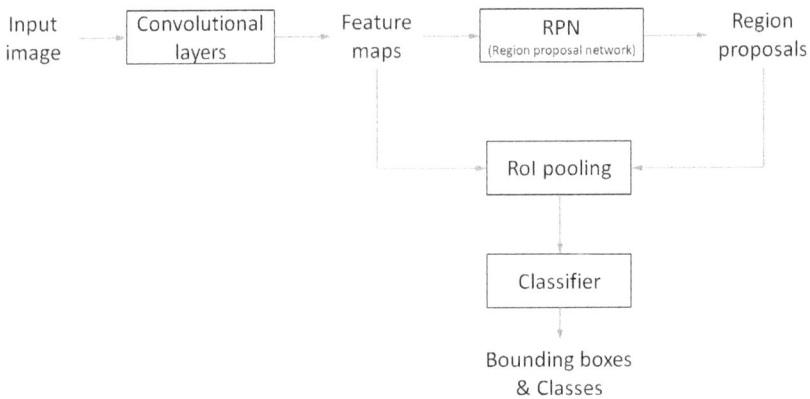

**Figure 1.8** The system framework of Faster R-CNN. (The figure is inspired by and drawn based on [182].)

upon this foundation by leveraging shared convolutional features across proposals, markedly accelerating the processing speed.

Faster R-CNN addresses the limitations of prior object detection methodologies by introducing a unified architecture that integrates both region proposal generation and object detection within a single network. This architecture consists of two main components: a Region Proposal Network (RPN) and a Fast R-CNN detector.

The RPN generates region proposals, which represent potential object bounding boxes, through the application of a compact network sliding over the convolutional feature map derived from the input image. These proposals are then refined and classified by the Fast R-CNN detector, which shares the convolutional layers with the RPN, enabling end-to-end training and faster inference (see Figure 1.8). By adopting this proposal-based approach, Faster R-CNN obviates the need for external region proposal methods like selective search, resulting in a more rapid and efficient object detection process.

The impact of Faster R-CNN on object detection is profound, as it notably enhances both the efficiency and precision of detection tasks. Through the elimination of external region proposal methods and the incorporation of proposal generation directly within the network, Faster R-CNN optimizes the object detection pipeline, yielding superior performance on benchmark datasets such as PASCAL VOC and COCO.

### 1.4.2.2 YOLO (You Only Look Once)

The YOLO network, introduced by Joseph Redmon et al. in their 2016 paper "You Only Look Once: Unified, Real-Time Object Detection," represents a pioneering approach to object detection [179]. Departing from conventional methods that necessitate multiple passes over an image to detect objects, YOLO employs a unified strategy by concurrently predicting bounding boxes and class probabilities in a single pass of the network. This streamlined architecture empowers YOLO to achieve real-time object detection with remarkable efficiency and accuracy.

The evolution of the YOLO algorithm has seen numerous iterations, from its inception as YOLOv1 to the latest version, YOLOv10 [221], each iteration striving to rectify limitations and enhance performance in terms of accuracy, speed, and robustness. YOLOv2 [180], introduced in 2017, incorporated enhancements such as batch normalization, high-resolution classifiers, and anchor boxes, which improved accuracy and stability. YOLOv3 [181], released in 2018, further refined the architecture by integrating feature pyramid networks (FPN) and multi-scale detection, resulting in heightened performance across diverse object sizes. YOLOv4 [13], unveiled in 2020, brought forth advanced techniques including a Cross-Stage Partial (CSP) backbone, Path Aggregation Network (PANet), and the Mish activation function. Subsequently, YOLOv5 [98] expanded on the innovations of YOLOv4, significantly improving detection speed and accuracy through innovations like Mosaic data augmentation, adaptive anchor box calculation, and Focus structure. YOLOv5 was released by the community, deviating from the original release model of the YOLO series. However, due to its excellent performance and user-friendly design, it gained rapid adoption among developers and researchers.

Overall, the YOLO series has significantly influenced the domain of object detection, offering effective and precise solutions for applications including surveillance, autonomous vehicles, and object tracking. Its innovative approach to real-time detection has stimulated subsequent research and advancements in the field, catalyzing the introduction of novel state-of-the-art detection models and methodologies.

### 1.4.3 Semantic Segmentation

Neural networks can segment images into semantically meaningful regions, assigning each pixel a class label to understand the spatial extent of objects within the image. This capability is essential for tasks requiring detailed scene understanding, such as autonomous driving, robotics, and satellite image analysis. Below are two classic networks that have played a milestone role in the field of semantic segmentation.

#### 1.4.3.1 *FCNs (Fully Convolutional Networks)*

Fully Convolutional Networks (FCNs) were introduced by Jonathan Long, Evan Shelhamer, and Trevor Darrell in their seminal paper "Fully Convolutional Networks for Semantic Segmentation," presented at the 2015 IEEE Conference on Computer Vision and Pattern Recognition (CVPR) [144]. The primary motivation behind FCNs was to adapt standard CNNs for dense prediction tasks, such as semantic segmentation, where the objective is to assign a class label to each pixel in an image.

Traditional CNNs typically employ fully connected layers at the end for classification tasks. In contrast, FCNs replace these fully connected layers with convolutional layers, thereby preserving spatial dimensions and accommodating variable input sizes. FCNs utilize deconvolutional layers (also known as transposed convolutional layers) to upsample feature maps back to the original input size, which is essential for generating pixelwise predictions (see Figure 1.9). Therefore, the FCNs support image input of any size and can produce correspondingly sized output. To recover fine details that may be lost during pooling and downsampling processes, FCNs incorporate

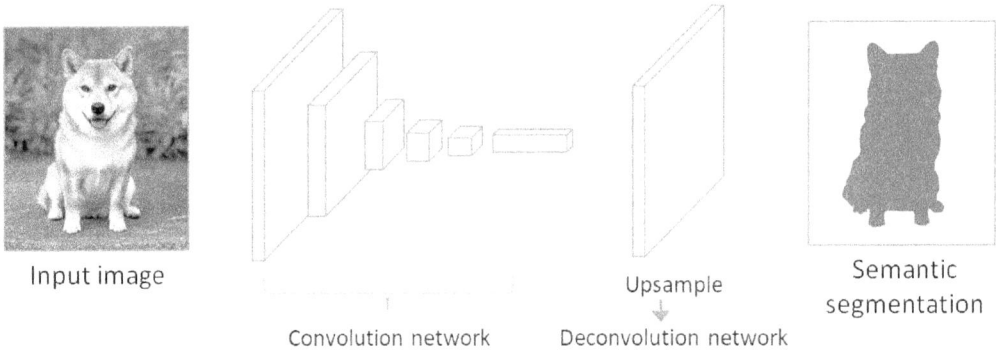

Input image      Upsample      Semantic segmentation

Convolution network      Deconvolution network

**Figure 1.9** A typical FCN architecture for semantic segmentation. This schematic illustration is a conceptual representation and does not depict a specific network structure. (The figure is inspired by and drawn based on [144].)

skip connections that combine coarse, high-level features with fine, low-level features from earlier network layers. This integration of multi-scale contextual information enhances segmentation accuracy. For more details, please refer to [144].

FCNs represent a significant breakthrough in semantic segmentation, fundamentally transforming the approach to dense prediction tasks in computer vision. FCNs enable end-to-end training for pixelwise predictions, greatly simplifying the semantic segmentation pipeline compared to previous methods that often required postprocessing steps. By leveraging the hierarchical feature extraction capabilities of CNNs and incorporating upsampling and skip connections, FCNs achieved state-of-the-art performance on several benchmarks, demonstrating the efficacy of deep learning for dense prediction tasks. Additionally, FCNs paved the way for a variety of dense prediction applications beyond semantic segmentation, including instance segmentation, depth estimation, and optical flow, thereby influencing a wide range of applications in computer vision.

### 1.4.3.2 U-Net

U-Net was proposed by Olaf Ronneberger, Philipp Fischer, and Thomas Brox in their paper "U-Net: Convolutional Networks for Biomedical Image Segmentation," presented at the International Conference on Medical Image Computing and Computer-Assisted Intervention (MICCAI) in 2015 [187]. The primary motivation behind U-Net was to address the challenges of biomedical image segmentation, where precise localization and efficient training with limited annotated data are crucial. Due to its exceptional segmentation performance, U-Net has been widely adopted in various domains of semantic segmentation, such as satellite image segmentation and so on.

U-Net's architecture is designed to capture both the context and precise localization needed for segmentation tasks (see Figure 1.10). It consists of two main parts: the contracting path (encoder) and the expansive path (decoder). The contracting path is similar to a typical CNN, comprising convolutional and max-pooling layers. The expansive path includes upsampling of the feature map followed by a 2x2 convolution

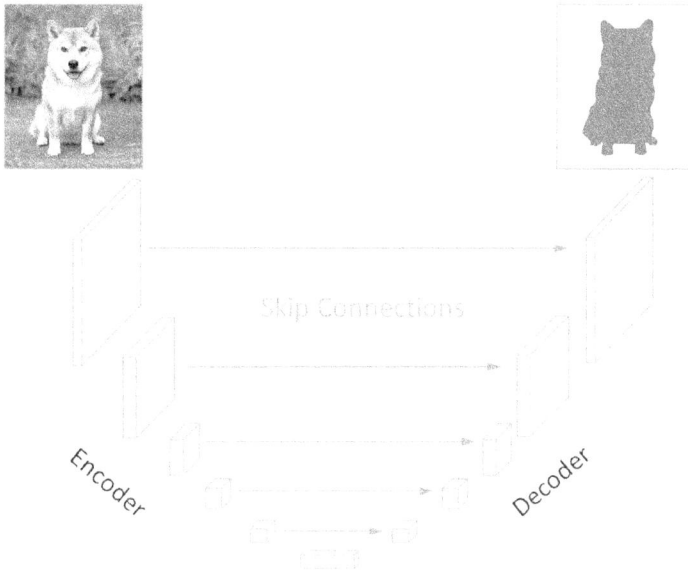

Figure 1.10 U-net architecture. This schematic illustration is a conceptual representation and does not depict a specific network structure. (The figure is inspired by and drawn based on [187].)

to halve the number of feature channels, concatenation with the corresponding high-resolution feature map from the contracting path, and two 3x3 convolutions each followed by a ReLU activation. Skip connections between the contracting and expansive paths concatenate feature maps from the encoder with corresponding layers in the decoder. These connections ensure the network retains high-resolution features that are essential for precise segmentation, effectively addressing the issue of losing fine details during downsampling. Finally, a 1x1 convolution is used at the final layer to map each feature vector to the desired number of classes.

U-Net has significantly advanced the field of image segmentation. First, U-Net can be trained with relatively small amounts of annotated data, which is especially beneficial in biomedical applications where large annotated datasets are often difficult to obtain. Second, the architecture's design, particularly the skip connections, ensures that high-resolution features are retained, resulting in precise and accurate segmentations. Third, variants such as U-Net++ [263] and Attention U-Net [158] have been developed to enhance performance and address specific challenges. Overall, the design principles of U-Net have inspired numerous advancements and applications across various domains.

### 1.4.4 Keypoint Detection

Keypoint detection, also known as keypoint localization, feature point detection, or landmark detection, is a crucial task in computer vision involving the identification of specific points of interest in an image. These keypoints are often used in applications such as object recognition, human pose estimation, facial landmark detection, and more. Traditional keypoint detection relies on hand-designed features and algorithms that use mathematical operations to identify specific patterns in images. Neural networks have significantly advanced keypoint detection by leveraging their ability to understand intricate patterns and relationships in image data, greatly improving accuracy and robustness. Below are two typical networks, which have been instrumental in this field.

### 1.4.4.1 OpenPose

OpenPose is an open-source library developed by the Carnegie Mellon Perceptual Computing Lab for real-time multi-person keypoint detection and pose estimation [21, 202, 20]. Initially released in 2017, OpenPose has since become one of the most popular tools for real-time multi-person keypoint detection due to its ability to accurately and efficiently detect human poses in images and videos, even in crowded scenes.

The OpenPose architecture consists of two main components: a feature extraction network and subsequent stages for predicting Part Affinity Fields (PAFs) and confidence maps (see Figure 1.11). Initially, a pre-trained CNN, like VGG-19, extracts feature maps from the input image. Next, a two-branch multi-stage CNN is employed. One branch predicts PAFs, which are 2D vector fields indicating limb direction and location. These predictions are refined through multiple stages, with each stage enhancing the PAFs by processing concatenated features from the previous stage and the original image features,

$$\begin{aligned} L^1 &= \varnothing^1(F) \\ L^t &= \varnothing^t(F, L^{t-1}), \forall\, 2 \le t \le T_P \end{aligned} \tag{1.1}$$

where $L^1$ refers to the PAF at stage 1, $\varnothing^1$ represents the CNNs for inference at stage 1, $L^t$ refers to the PAF at stage $t$, $\varnothing^t$ represents the CNNs for inference at stage t, $F$ denotes the feature maps, and $T_P$ signifies the total number of PAF stages.

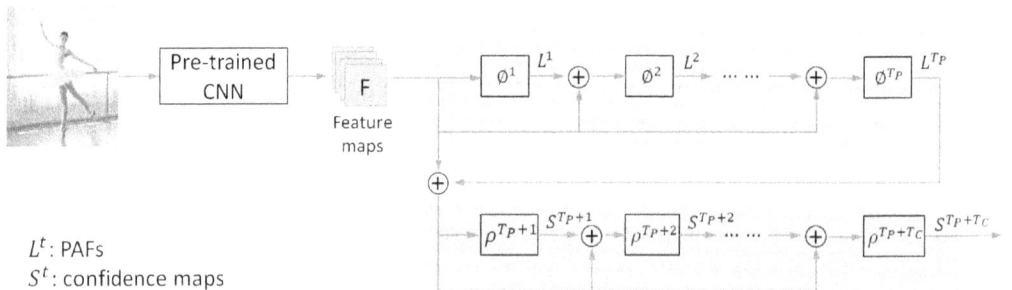

$L^t$: PAFs
$S^t$: confidence maps

**Figure 1.11** The architecture of OpenPose. This schematic illustration is a conceptual representation and does not depict a specific network structure. (This figure is inspired by and drawn based on [20].)

Following this, the other branch predicts confidence maps for body part locations, indicating the probability of each pixel representing a specific body part. Like PAFs, confidence maps are iteratively refined,

$$S^{T_P} = \rho^{T_P}(F, L^{T_P})$$
$$S^t = \rho^t(F, L^{T_P}, S^{t-1}), \forall T_P \leq t \leq T_P + T_C$$

(1.2)

where $S^{T_P}$ refers to the confidence map at stage $T_P$, $\rho^{T_P}$ represents the CNNs for inference at stage $T_P$, $S^t$ refers to the confidence map at stage $t$, $\rho^t$ represents the CNNs for inference at stage $t$, and $T_C$ signifies the total number of confidence map stages.

The final outputs, consisting of refined PAFs and confidence maps, go through bipartite matching for parsing. This step associates detected body parts and assembles them into complete body poses for all individuals in the image.

OpenPose is one of the first frameworks capable of performing real-time multi-person pose estimation with high accuracy. One of its key contributions is the introduction of PAFs, which model the spatial relationships between body parts, enabling precise and robust limb detection even in crowded scenes. By employing a multi-stage CNN architecture, OpenPose iteratively refines its predictions, enhancing accuracy at each stage. Its open-source nature has facilitated widespread adoption and further innovation, impacting various fields such as healthcare, animation, and surveillance. Additionally, OpenPose's ability to function with relatively low computational requirements has made advanced pose estimation accessible to a broader range of researchers and developers, spurring growth and development in computer vision technologies.

### 1.4.4.2 HRNet (High-Resolution Network)

HRNet, or High-Resolution Network, was proposed by Jingdong Wang and his colleagues in their 2019 paper "Deep High-Resolution Representation Learning for Visual Recognition" [228]. The network aims to overcome the limitations of traditional CNNs in maintaining high-resolution representations throughout the architecture, which is critical for position-sensitive vision problems, such as human pose estimation [208].

Unlike conventional networks that progressively downsample feature map resolutions, HRNet maintains multiple parallel streams of varying resolutions (see Figure 1.12). It initiates with a high-resolution stream and incrementally introduces lower-resolution streams in parallel. HRNet executes repeated multi-scale fusions to facilitate information exchange across these different resolution streams. These fusions merge high-resolution and low-resolution representations, allowing the network to simultaneously utilize detailed spatial information and robust semantic information. The architecture is organized into stages, each comprising several residual blocks. These blocks independently process feature maps within each stream before the multi-scale fusion steps integrate the information across different streams. In the final stage, all parallel streams are consolidated to generate a high-resolution representation. This

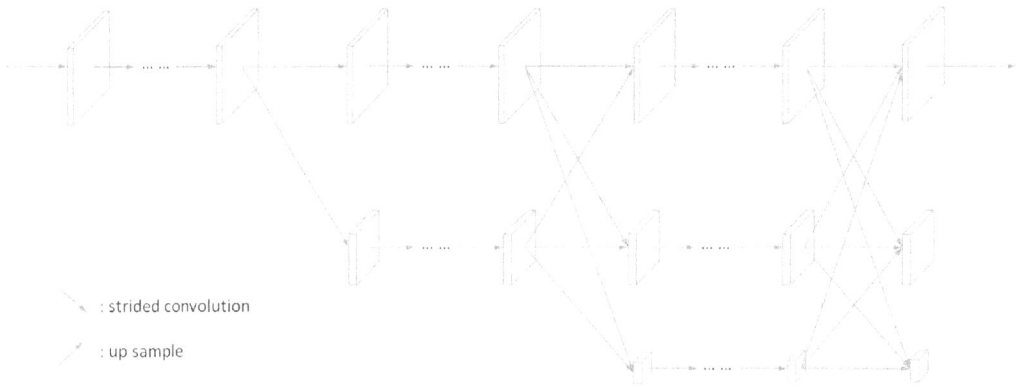

: strided convolution

: up sample

**Figure 1.12** Network structure illustration of HRNet. (This figure is inspired by and drawn based on [208].)

high-resolution representation is then employed for various downstream tasks, yielding detailed and accurate predictions.

HRNet signifies a major breakthrough in neural network architecture by preserving high-resolution representations via parallel high-resolution streams and multiscale fusion. This approach circumvents the spatial detail loss typically associated with traditional downsampling methods, which is vital for tasks such as pose estimation and semantic segmentation that require fine spatial details. HRNet has demonstrated state-of-the-art performance across multiple benchmarks. Its robustness and accuracy have facilitated its integration into practical applications, including autonomous driving, medical imaging, and video surveillance, where precise spatial localization is paramount.

### 1.4.5 Image Generation

Image generation is an important branch in the field of computer vision. It can generate realistic images from random noise and plays an important role in image processing, such as image denoising, image restoration, etc. The following introduces two classic image generation networks.

#### 1.4.5.1 GANs (Generative Adversarial Networks)

The Generative Adversarial Network (GAN) was introduced by Ian Goodfellow and his colleagues in 2014 [70]. It consists of two primary components: the generator and the discriminator. The generator takes random noise as input and aims to produce samples that resemble real data. The discriminator, on the other hand, receives both the generated samples and real data, and its task is to distinguish between the two. During the training process, the generator and discriminator engage in a competition, each improving its performance. The generator refines its ability to create realistic samples, while the discriminator becomes better at identifying them. This iterative process continues until the generator is able to produce sufficiently realistic images that can deceive the discriminator.

Essentially, GAN is designed to train a generative model that aims to produce data following the same distribution as real data. Since the real data distribution is unknown, GAN employs an adversarial approach, where the training of the generative model is guided by a classifier–the discriminator. Let the distribution of the generator over the data $x$ be $p_g$, and define $p_z(z)$ as the prior distribution of the input noise variable. The generator, parameterized by $\theta_g$, maps the input noise $z$ to a data space as $G(z; \theta_g)$. The discriminator $D(x, \theta_d)$ parameterized by $\theta_g$, outputs a scalar, which represents the probability that $x$ comes from the real data rather than from $p_g$. The discriminator is trained to maximize the probability of correctly labeling real data and generated data, while the generator is trained to minimize $log(1 - D(G(Z)))$. This adversarial process can be framed as a two-player minimax game with the following value function $V(G, D)$:

$$\min_{G} \max_{D} V(D, G) = \mathbb{E}_{x \sim p_{data}(x)}[\log D(x)] + \mathbb{E}_{z \sim p_z(z)}[\log(1 - D(G(z)))] \qquad (1.3)$$

After the proposal of the GAN, a series of classic GANs emerged, such as DCGAN [173], InfoGAN [30], WGAN [7], CycleGAN [265], BigGAN [15], StyleGAN [105, 106], etc. The Deep Convolutional Generative Adversarial Network (DCGAN), proposed by Radford et al. in 2015, is a significant variant of GAN that replaces the fully connected layers in traditional GANs with convolutional layers, thereby enhancing the quality and stability of generated images [173]. InfoGAN introduces interpretability to the latent space by maximizing the mutual information between the latent code and the generated samples [30]. The Wasserstein GAN (WGAN) improves the training stability of GANs by leveraging the Wasserstein distance as a more effective loss function [7]. Another notable variant, Cycle-Consistent Generative Adversarial Network (CycleGAN), proposed by Zhu et al. in 2017, enables unpaired image-to-image translation by introducing a cycle-consistency loss [265]. BigGAN scales up GAN architectures by increasing network parameters and batch sizes, exploring the effects of such scaling and addressing associated instabilities [15]. StyleGAN, introduced by Karras et al. in 2018, marked a significant advancement in GANs by incorporating style embeddings and adaptive instance normalization (AdaIN), enabling independent control over the style and content of generated images. StyleGAN2, an improved version, further refined performance by addressing key limitations of the original architecture and enhancing the quality of generated images [106]. StyleGAN excels in tasks such as high-resolution image generation, image editing, and style transfer, producing highly realistic and diverse outputs.

GANs represent a transformative framework in the field of image generation, renowned for their capacity to produce high-quality, realistic images through an adversarial training process between a generator and a discriminator. In addition to image generation, GANs excel in tasks such as style transfer, super-resolution, and image restoration, showcasing their versatility across diverse applications. Despite these successes, GANs face significant challenges, including training instability, mode collapse, and the complexity of hyperparameter tuning. These limitations have spurred the exploration of alternative methods, such as diffusion models, which aim to address GANs' shortcomings. Nevertheless, GANs have played a pivotal role in advancing generative modeling and remain a cornerstone of computer vision research.

### 1.4.5.2 Diffusion Models

The diffusion model is a generative framework grounded in probability theory, originally inspired by diffusion processes in physics, such as the dispersion of ink in water. In the context of machine learning, this concept has been innovatively adapted to data generation tasks, particularly in synthesizing images and sounds. Diffusion models simulate a gradual transition from complex data distributions to simple noise distributions and subsequently reconstruct high-quality data samples from noise by learning the inverse process.

The foundational work on diffusion models, titled "Deep Unsupervised Learning Using Nonequilibrium Thermodynamics," was published in 2015 [205]. This paper introduced a Markov chain-based framework for transforming one distribution into another, enabling the transformation from noise images to target images. In 2020, the advent of Denoising Diffusion Probabilistic Models (DDPM) marked a pivotal moment in the use of diffusion models for image generation [81]. DDPM applied the "denoising" diffusion probabilistic framework to high-quality image synthesis, establishing diffusion models as a mainstream approach in the field of image generation. Figure 1.13 shows the directed graphical model of DDPM.

The fundamental concept of the diffusion model can be divided into two distinct processes: the forward diffusion process and the reverse generation process. In the forward diffusion process, starting from a real sample data $\mathbf{x}_0$, noise is incrementally added, producing a sequence of intermediate states $\mathbf{x}_1$, $\mathbf{x}_2$, ..., $\mathbf{x}_T$, until the final state $\mathbf{x}_T$ approximates a standard Gaussian distribution. At each time step $t$, the addition of a small amount of Gaussian noise progressively blurs the data. The forward diffusion process can be mathematically expressed as follows:

$$q(\mathbf{x}_t|\mathbf{x}_{t-1}) = \mathcal{N}(\mathbf{x}_t; \sqrt{1-\beta_t}\mathbf{x}_{t-1}, \beta_t\mathbf{I}) \tag{1.4}$$

where $\beta_t$ is the noise intensity at each step, $\mathcal{N}(\mu, \Sigma)$ denotes a Gaussian distribution with mean $\mu$ and covariance $\Sigma$, and $\mathbf{x}_t$ is the intermediate state in the diffusion process.

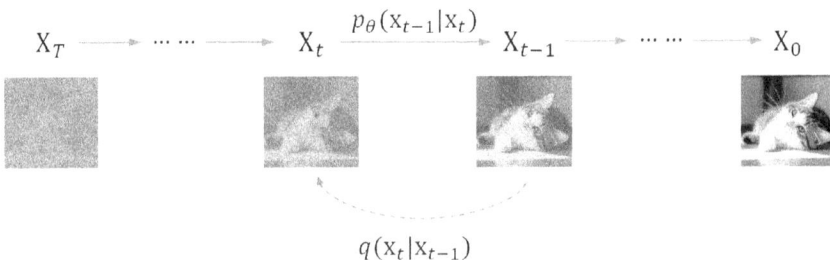

Figure 1.13   The directed graphical model of DDPM. (This figure is inspired by and drawn based on [81].)

The forward process over all steps can be expressed as:

$$q(\mathbf{x}_{1:T}|\mathbf{x}_0) = \prod_{t=1}^{T} q(\mathbf{x}_t|\mathbf{x}_{t-1}) \tag{1.5}$$

The backward generation process reconstructs data by reversing the forward process, starting from a Gaussian noise sample $\mathbf{x}_T$ and sequentially denoising it to obtain $\mathbf{x}_0$. Each step of the reverse process is approximated by a learned neural network model $p_\theta(\mathbf{x}_{t-1}|\mathbf{x}_t)$, whose goal is to reverse the process of adding noise as accurately as possible:

$$p_\theta(\mathbf{x}_{t-1}|\mathbf{x}_t) = \mathcal{N}(\mathbf{x}_{t-1}; \mu_\theta(\mathbf{x}_t, t), \Sigma_\theta(\mathbf{x}_t, t)) \tag{1.6}$$

where $\mu_\theta(\mathbf{x}_t, t)$ is the predicted mean at step $t$, $\Sigma_\theta(\mathbf{x}_t, t)$ is the predicted variance.

The joint probability of the reverse process over all steps is:

$$p_\theta(\mathbf{x}_{0:T}) = p(\mathbf{x}_T) \prod_{t=1}^{T} p_\theta(\mathbf{x}_{t-1}|\mathbf{x}_t) \tag{1.7}$$

Diffusion models have emerged as a powerful framework for image generation, offering several advantages over traditional generative models. These models are capable of producing highly realistic and detailed images, often surpassing other methods, such as GANs, particularly on complex datasets. Their probabilistic formulation ensures stable training, addressing common issues like mode collapse and instability found in GANs. Diffusion models are also robust to hyperparameter variations, providing better coverage of diverse data distributions. Furthermore, their flexibility allows adaptation to various tasks, including conditional generation, by integrating auxiliary information such as text or class labels. They scale effectively to high-resolution datasets, maintaining both quality and diversity, making them suitable for a wide range of applications.

The applications of diffusion models span numerous domains in image generation. They excel in unconditional image synthesis, generating high-quality images from random noise, as well as in conditional tasks such as image-to-image translation and content generation guided by textual or semantic inputs. In the field of image restoration, diffusion models are proficient in tasks like denoising, super-resolution, and inpainting, demonstrating their ability to recover degraded images with remarkable fidelity. Moreover, these models facilitate creative endeavors, including artistic image generation and style transfer, offering innovative tools for visual arts and design.

Beyond image generation, diffusion models have been extended to the domain of video generation, addressing the challenges associated with producing coherent and realistic sequences over time. By modeling the temporal dependencies alongside spatial information, these models can generate high-quality video sequences from noise or conditioned inputs, such as prior frames or textual descriptions. This capability has opened new possibilities in animation, video restoration, and multimedia content creation, marking diffusion models as a significant advancement in video synthesis.

Several key models illustrate the progression of this technology. DDPM introduced a probabilistic framework for iterative denoising, establishing diffusion models as a mainstream approach to high-quality image generation [81]. Improved DDPM enhanced generation quality through optimized noise schedules and architectures, often surpassing GAN-based methods [155]. Score-based Generative Model advanced the framework by learning the score function of data distributions and integrating it with the diffusion process for better synthesis quality [207]. Latent Diffusion Models (LDM) improved computational efficiency by working in a reduced-dimensional latent space while maintaining output fidelity [186]. Advanced models like Imagen and DALL·E 2 have further pushed the boundaries of text-to-image generation, demonstrating the versatility and power of diffusion models.

The diffusion model has become a leading framework in both image and video generation, gaining widespread recognition for its ability to produce high-quality results. However, it faces notable challenges, including high computational costs and slow sampling speeds. With advancements in algorithmic optimization and hardware technology, these limitations are expected to be addressed, leading to further improvements in both the efficiency and quality of generated outputs.

## 1.5   SHORTCOMINGS OF THE EXISTING SYSTEMS

Existing vision systems based on neural networks have made remarkable progress across various applications. However, they still face significant challenges that hinder their widespread adoption and effectiveness. Three primary issues stand out: high data dependence, poor interpretability, and vulnerability to adversarial attacks.

**(1) High Data Dependence**

Neural networks rely on substantial amounts of annotated data for training to attain high accuracy and generalizability. This poses a significant challenge in domains where acquiring and labeling large datasets is difficult, time-consuming, or costly. This reliance on data is evident in two key aspects:

- Training Requirements: Neural networks, especially deep learning models, necessitate extensive annotated data to effectively learn numerous parameters, which helps the model generalize well to unseen data. However, obtaining such large, high-quality datasets is often arduous, particularly in specialized domains like medical imaging, where expertise and substantial resources are indispensable for data acquisition and labeling.

- Generalization Issues: Models trained on limited or biased datasets may encounter difficulties in adequately generalizing to new, unseen data due to the lack of diversity in the training data. This phenomenon, known as overfitting, results in the model performing well on the training data but inadequately on real-world applications.

**(2) Poor Interpretability**

Neural networks often operate as "black boxes," presenting challenges in understanding their decision-making processes and predictions. This lack of interpretability undermines trust and confidence in the model's outputs.

- Black-Box Nature: Neural networks, particularly deep networks, function as intricate black boxes, lacking transparency in their internal workings and decision-making mechanisms. This opacity makes it challenging to elucidate how specific outputs are derived from given inputs.

- Lack of Insight: This opacity hampers our ability to gain insights into the model's functionality, diagnose errors, or understand the features driving predictions. In critical domains like autonomous driving or healthcare, the inability to interpret decisions can erode trust and impede model adoption.

- Model Debugging and Improvement: Poor interpretability further complicates the debugging process. Identifying and rectifying specific internal features or weights causing errors in predictions becomes arduous when the model operates as a black box.

### (3) Vulnerability to Adversarial Attacks

Neural networks are vulnerable to adversarial attacks, wherein minor, deliberately crafted alterations to input data can result in substantial misclassifications or erroneous outputs. This susceptibility underscores concerns regarding the robustness and security of neural network-based visual systems, particularly in critical applications.

The following describes several instances of successful adversarial attacks using adversarial samples. In one example, a study by researchers from KU Leuven in Belgium demonstrated that a simple printed pattern, such as a sticker, can effectively bypass AI video surveillance systems by evading detection in human body recognition [213]. Additionally, their research introduced an "adversarial patch," a piece of colored cardboard that tricks the AI system, preventing it from detecting individuals in the scene.

Another example involves an algorithm developed by Parham Aarabi, a professor at the University of Toronto, and his graduate student Avishek Bose [14]. This algorithm generates small perturbations through training an adversarial generative network, causing pre-trained face detectors to fail when the perturbations are applied to input face images.

Further research by the Real AI team at the Institute of Artificial Intelligence of Tsinghua University demonstrated how AI algorithms could produce specific patterns to bypass facial recognition systems. Their approach successfully compromised facial unlocking systems on 19 mobile phones and breached several financial and government service applications.

Attacks on autonomous driving systems have also been demonstrated. For instance, research by Keen Security Lab et al. successfully misled the lane detection module of Tesla's Autopilot system by introducing small, imperceptible perturbations to the camera image, causing the vehicle to follow false lane markings into the opposite lane [97]. Similarly, work by Kevin Eykholt et al., presented in their CVPR 2018 paper, showed that attaching black and white stickers to physical road signs could cause a visual classifier to misinterpret a stop sign as a Speed Limit 45 sign [55]. These examples highlight the vulnerabilities of AI systems to adversarial manipulation.

## 1.6 ROOT CAUSES OF THE CHALLENGES

We believe that the shortcomings of deep learning systems with deep neural networks at their core mainly stem from two fundamental reasons:

**(1) Inherent Limitations of Purely Data-Driven Methods:** These methods are highly dependent on the integrity and representativeness of training data. However, real-world data is complex and variable, making it difficult for training datasets to fully cover all practical scenarios. Consequently, when encountering new problems not represented in the training data, deep learning models may produce unexpected and uncontrollable predictions. This unpredictability challenges system stability and contributes to the vulnerability of deep learning models.

**(2) Intrinsic Characteristics of Neural Network Architecture:** Neural networks operate as "black box" models, with complex internal mechanisms that are difficult to explain intuitively, leading to a lack of model interpretability. Current research indicates that some key features of neural networks contribute to their vulnerability to attacks. For example, the learned mapping from input to output often exhibits a high degree of discontinuity [211], and the linear behavior of deep neural networks in high-dimensional spaces provides opportunities for attackers [72].

In summary, the limitations of purely data-driven approaches and the inherent characteristics of neural network architectures collectively form the core roots of the challenges faced by deep learning systems.

## 1.7 PRACTICAL IMPLEMENTATION STRATEGIES

### 1.7.1 Optimizing Usability

To make deep learning models more practical, the following key strategies can be employed (see Figure 1.14):

**(1) Algorithm Innovation:** Continuously enhance the core mechanisms of the algorithm, including but not limited to the optimization of learning algorithms, the innovation of learning mechanisms, and the integration of artificial rules. This helps the model learn deeper and more essential features from the data, improving its generalization and accuracy.

**(2) Scenario Verticalization:** Focus the model's practical applications on specific, well-defined scenarios that closely match the conditions under which the model

Figure 1.14 Strategies for optimizing usability.

was trained. This reduces uncertainty in complex environments and improves the model's performance on specific tasks.

(3) **Data Augmentation and Generation:** Expand the training dataset by collecting and annotating more data or using data augmentation techniques to create diverse samples. A rich dataset exposes the model to a wider range of patterns, enhancing its adaptability and robustness.

(4) **Growth Mechanism:** Design the model to self-learn and improve in real-world applications through mechanisms like online learning and incremental learning. This allows the model to continuously adjust and optimize based on feedback and new data in practical use, achieving ongoing performance improvements rather than relying solely on initial training.

(5) **Enhanced Multi-modal Integration:** Equip the model with the ability to handle multiple tasks and integrate data from various sources and types (e.g., images, text, sound). This capability enables the model to establish connections between different modalities, driving cross-domain understanding and reasoning, and advancing deep learning towards more general and intelligent applications.

## 1.7.2 Industry AI Paradigm

Human intelligent behavior and its developmental patterns always offer profound insights into the development of AI. Upon observation, it becomes evident that, despite sharing similar brain structures, sensory organs, and motor functions, individuals can cultivate a wide array of specialized skills throughout their unique life paths. These skills are not static or fixed, but show a dynamic evolutionary process, accompanied by the continuous learning and growth of individuals, as shown in Figure 1.15.

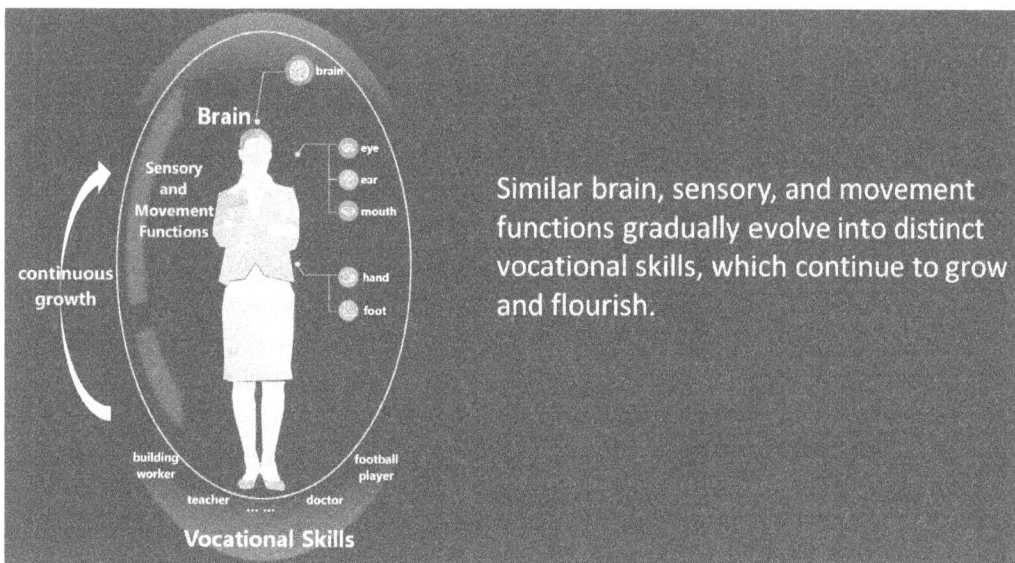

Figure 1.15   The growth model of human capabilities.

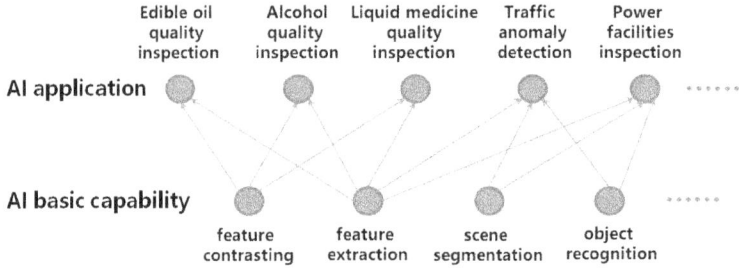

Figure 1.16  Traditional construction method for AI applications.

Inspired by the growth model of human capabilities and combining extensive project practices, we have developed an industry AI paradigm designed to address the challenges posed by fragmented industry scenarios. Traditionally, each application scenario requires customization of a set of AI models, a process that is both costly and time-consuming (see Figure 1.16).

To overcome this fragmentation, we have systematically identified and summarized the commonalities across various scenarios, resulting in the creation of an industry AI paradigm (see Figure 1.17).

This paradigm functions as a versatile tool that can solve problems across multiple similar scenarios, enabling rapid adaptation to new application contexts while minimizing the need for extensive customization. For example, the liquid quality detection paradigm, developed from scenarios such as edible oil quality detection, alcohol quality detection, and liquid medicine quality detection, can be quickly adapted to address the challenge of liquid cosmetics quality detection. This approach streamlines the deployment of AI solutions across industries, significantly reducing development time and cost.

### 1.7.3  AI Engineering Framework

Expanding on the paradigm, a comprehensive AI engineering framework has been developed, as shown in Figure 1.18. At its foundation are computing-network con-

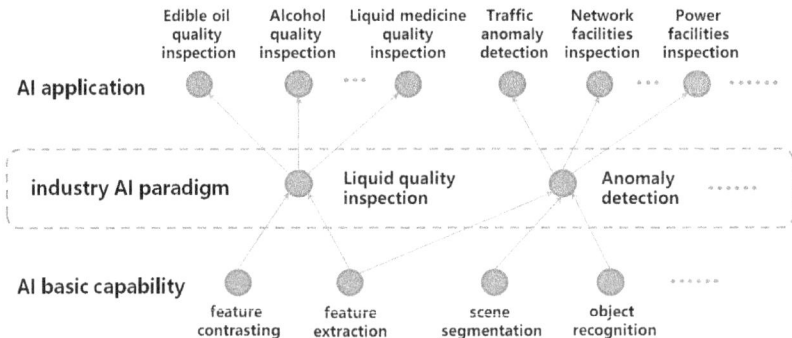

Figure 1.17  Construction method based on industry AI paradigm.

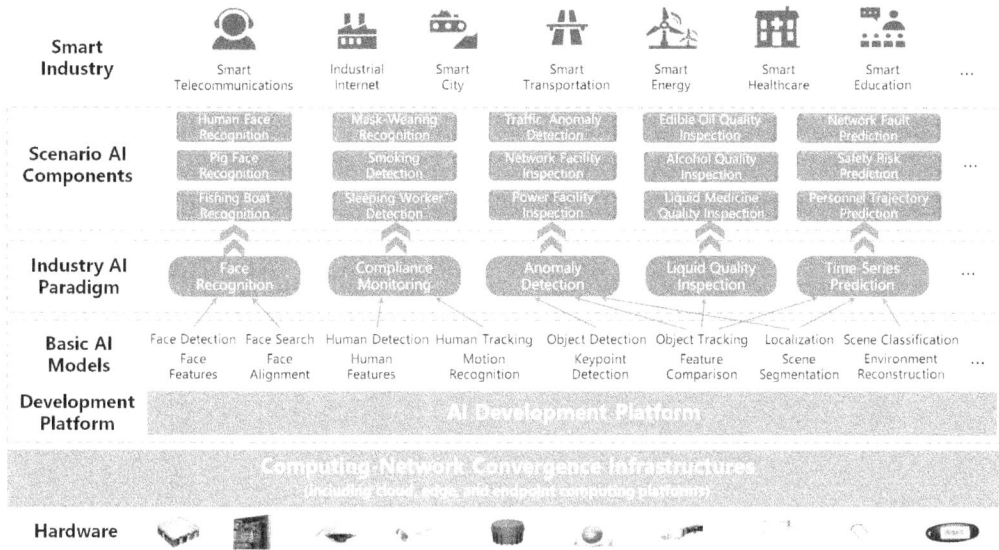

**Figure 1.18** AI engineering framework based on industry AI paradigm.

vergence infrastructures and an advanced AI development platform. The base layer comprises a wide array of AI foundational models, including face detection, face feature extraction, face retrieval, face alignment, human body detection, human body tracking, human body feature extraction, motion recognition, object detection, and more. These foundational models serve as the "molecular capabilities" for constructing complex AI systems.

Building upon these foundational models, the framework establishes a series of industry AI paradigms by identifying and summarizing the commonalities across various application scenarios. These paradigms encapsulate best practices and proven algorithms for addressing recurring challenges, enabling efficient reuse and rapid problem-solving in similar contexts. By formalizing these paradigms, the framework significantly reduces the time required to move from concept to implementation, accelerating the adoption and integration of AI technologies across diverse industries.

On top of the industry AI paradigms, the framework supports the development of scenario-specific components tailored to the unique requirements of particular industries. These components are deeply customized by incorporating the business processes, data characteristics, and user needs of specific scenarios. They leverage the industry AI paradigms as a foundation, further refining and optimizing solutions. This modular approach simplifies the construction of more intricate systems, providing flexibility and scalability.

Ultimately, by seamlessly integrating these diverse components across industries and application scenarios, the framework enables the creation of sophisticated and powerful smart industry systems.

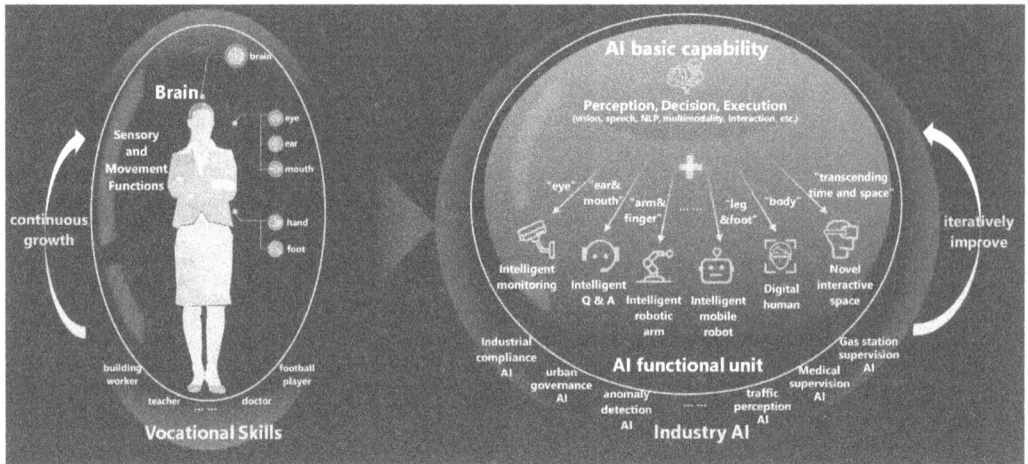

Figure 1.19   AI development model similar to the growth model of human capabilities.

### 1.7.4   AI Development Model

Corresponding to the growth model of human capabilities, the development model of AI reflects a similar trajectory (see Figure 1.19). This model begins with the creation of a foundational AI capability library, encompassing three core domains: perception, decision-making, and execution. This library includes capabilities such as visual perception, language processing, natural language understanding, multimodal comprehension, and interactive functionality.

Based on this foundation, AI functional units are developed to emulate human faculties. These units include intelligent monitoring systems, conversational AI, robotic arms, mobile robots, and digital humans, which correspond to human sensory organs and physical components, such as eyes, ears, limbs, fingers, feet, and the body. Together, the capability library and functional units establish a versatile and robust general-purpose foundation, comparable to a physically capable individual.

From this general foundation, these capabilities are customized to meet the specific needs of various industries through deep specialization. This process results in the development of industry-specific AI solutions, akin to professional skills tailored for different fields. Examples of these solutions include AI systems for industrial compliance, urban governance, anomaly detection, traffic perception, gas station supervision, and medical regulation. These specialized AI systems enable diverse industries to achieve greater efficiency and innovation.

Like human expertise, these industry-specific AI systems undergo continuous improvement through iterative application and refinement. Over time, they reach higher levels of sophistication and performance, further enhancing their capacity to address complex, real-world challenges and make a significant impact across various sectors.

In summary, the AI engineering framework illustrated in Figure 1.18 presents a systematic and modular approach to building intelligent industries. Complementing this, the AI development model depicted in Figure 1.19 serves as the conceptual foundation and source of inspiration for the framework. By integrating foundational AI models, refining industry-specific AI paradigms, and tailoring scenario-specific components, this approach has successfully driven the widespread adoption and seamless integration of AI technologies across diverse industries. The framework not only enhanced operational efficiency but also advanced the development of smarter, more adaptive systems capable of addressing complex challenges and meeting evolving demands.

# CHAPTER 2

# Fine-Grained Dataset Design Paradigm for Mask-Wearing Recognition

## 2.1 INTRODUCTION

With the development of deep learning [117, 194, 69, 47, 93, 185, 149], computer vision algorithms have been widely used in industrial scenarios. To achieve better results, it is necessary to provide large-scale relevant datasets. However, obtaining a significant amount of industrial scenario data is exceedingly challenging. In academic circles, pre-training on large-scale open-source datasets still reigns supreme. To accomplish associated tasks, better networks and parameters are then fine-tuned on smaller real-world scenario datasets. But the approach may bring some problems, of which we take object detection task as example for explanation. VOC [54, 52] and COCO [133, 31, 166] are the most commonly used open source datasets for model pre-training in the field of object detection. But for industrial application scenarios, the data division of VOC and COCO is too coarse. Taking person-counting task as an example, COCO labels parts of human body (such as fingers, single leg) as person. In complex industrial scenarios, such kind of data division method can easily lead to objects similar to legs or fingers being falsely detected as human, resulting in incorrect counts.

Figure 2.1 shows the feature paradigm of mask-wearing recognition dataset, with distinguishable samples at two ends and indistinguishable samples in the center. Difficulty level of recognition gradually increases from the two ends to the center, and the samples closer to the center are more likely to be misidentified. Many datasets for industrial applications fit this paradigm as well, such as helmet detection, work uniform recognition, glove wearing recognition, etc. In industrial application scenarios, we designate the categories that need extra attention as positive samples, such as not wearing mask in mask recognition, not wearing helmet in helmet detection, not wearing gloves in glove recognition, etc. Positive samples are closely related to management and security in practical applications, and precision rate is more important than recall rate, thus, we need to ensure the recognition precision of positive samples.

DOI: 10.1201/9781003644972-2

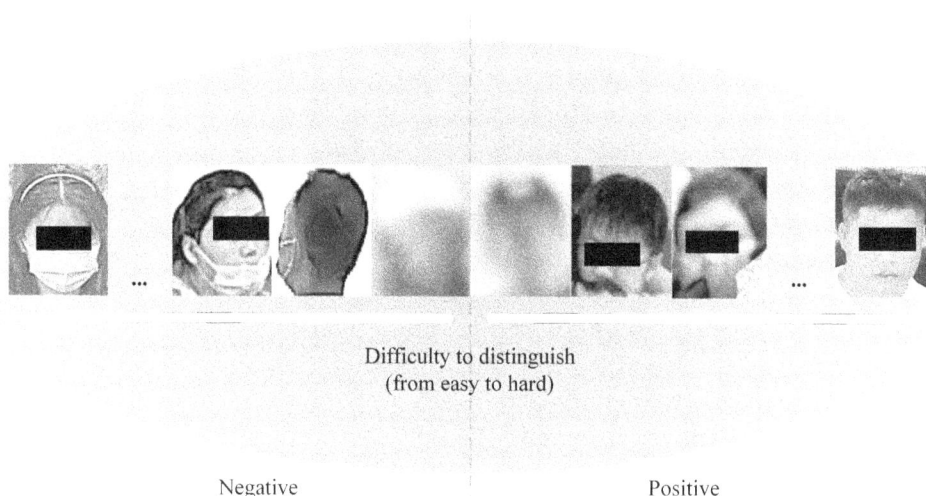

Figure 2.1   The feature paradigm of mask-wearing recognition dataset, with distinguishable samples at two ends and indistinguishable samples in the center. Difficulty level of recognition gradually increases from the two ends to the center, and the samples closer to the center are more likely to be misidentified.

When the model misidentifies negative samples as positive ones, false alarm occurs. Frequent false alarms will bring very terrible user experience. Therefore, reducing false alarms is a big challenge we need to address.

Currently mainstream approach of dataset design methods ignore the characteristics of the data itself and the actual scenarios. Take mask-wearing recognition as an example, according to the mainstream approach of dataset design, the data will be conservatively assigned to positive and negative sample sets. When the data scale is not large enough, the model trained in this way is disastrous. Because the actual application scenarios are very complex, i.e., irregularly wearing(e.g., a part of the mouth or nose is exposed), large angles and large poses, low quality, small targets, etc., model trained on these samples may produce a large number of false alarm(FA). Here FA stands for misidentifying mask-wearing face or other non-face object as not wearing mask. Therefore, we need to finely design the dataset according to the actual application scenarios and the own characteristics of dataset itself.

In this chapter, we finely design the dataset driven by application, based on the data-centric AI paradigm. By mining the own features of data and flexibly selecting the positive and negative sample sets according to the actual requirements, and adding the remaining samples to the training set as uncertainty classes, models trained on the dataset generated in proposed way can significantly reduce FA.

## 2.2   RELATED WORK

Large-scale datasets with high-quality annotations play a crucial role in driving better computer vision models. For image classification, the most commonly used datasets are ImageNet [42, 112, 200, 193]. ImageNet is a large image dataset built to facilitate the development of image recognition technology, containing 14,197,122 images and 21,841 categories. For object detection, PASCAL VOC [54, 52] is an early benchmark that contains 17,000 images in 20 categories. Then there is COCO [133, 31, 166] in 2014, which is currently the most widely adopted benchmark for object detection. It contains 118,000 images and 860,000 instance annotations in 80 categories.

ImageNet and COCO datasets with the characteristics of large-scale and high-quality together with deep learning have revolutionized the face of computer vision (CV). Most new algorithms assess their performance on these datasets and provide pre-trained models. We can reuse these pre-trained models in similar domains through transfer learning [162, 266, 156, 90]. The advantage of transfer learning is that it can produce excellent results with little in the way of training time, training data, or computational resources. Therefore, transfer learning is often the first choice in research and industry when developing deep learning-based models for CV tasks. However, pre-trained models may carry inherent hazards for domains with unique business needs. Data division of open source datasets is coarse and inaccurate for actual application scenarios, which makes the transfer-learned model incapable of distinguishing difficult samples in real scenarios.

Andrew Ng proposes data-centric AI [154] to address the above problems by designing targeted subsets of data. Data-centric AI is the discipline of systematically engineering the data needed to successfully build an AI system. For many practical applications, it is more efficient to improve the data than the network structure. Meanwhile, since many businesses simply lack enormous data volumes, the focus has to shift from big data to good data.

We illustrate our data design approach with mask-wearing recognition task. COVID-19 has made a huge impact on our lives. Wearing mask is the most effective way to protect against COVID-19. AIZOO [32] provides the face mask detection dataset which includes parts of WIDER Face [241] with Masked Faces (MAFA) datasets [60]. In AIZOO, training set contains 6,120 images and 13,593 faces, and the test set has 1830 images and 5082 faces. The Moxa 3K dataset created by Roy et al [191] is used for training and evaluating face mask detection models, which consists of 3,000 images with different scenarios from close-up faces to crowded scenes, using 2,800 images for training and 200 images for testing. However, the above-mentioned datasets are only simply divided into two categories: masked and unmasked, without considering the characteristics of the data and the needs of the application scenarios, leading to a lot of FA. In our work, we adopt the data-centric AI and carefully design the mask-wearing dataset, which effectively reduces the FA of the model.

## 2.3   DATASET AND FRAMEWORK

In this section, we introduce the proposed approach with the example of mask wearing recognition, including two parts: data collection and design.

## 2.3.1  Image Collection

We collect data from 75 real-world application scenarios including hospitals, health clinics, neighborhoods, schools, streets, offices, gas stations, factories, cold chains, warehouses, etc., and extract head parts in each scenario using existing head detectors, and finally build more than 10,000 images of mask-wearing recognition data. The dataset covers both masked and unmasked human head images with various angles, illumination, sizes, and quality. It is manually labeled by a number of people in numerous rounds, and it is divided into positive and negative subsets based on the mainstream dataset construction.

## 2.3.2  Data Design

The overall architecture of our dataset design scheme is presented in Figure 2.2. The current mainstream dataset division method is described in Figure 2.2 (a), which conservatively divides the dataset into positive and negative categories. Figure 2.2 (b) extracts the samples with irregular wearing from Figure 2.2 (a), Figure 2.2 (c) extracts the low-quality samples, and Figure 2.2 (d) extracts the samples with mask-like occlusion. In the proposed dataset designing scheme, we select the unambiguously non-masked and masked samples as positive and negative sets as depicted in Figure 2.2 (a) and Figure 2.2 (b), which is based on the characteristics of the data and the requirements of the application scenario. The samples that are easy to cause FA, such as irregular wearing, low-quality, mask-like occlusion, etc., are selected as the uncertain class to form the final training set.

### 2.3.2.1  Straightforward Dataset Design

Straightforward dataset design refers to directly dividing the dataset into positive and negative categories, as shown in Figure 2.2 (a), which is the current mainstream method. In mask-wearing recognition, due to the indistinguishable samples such as irregularly wearing, low-quality, mask-like occlusion etc., model trained on the dataset designed in a straightforward manner are disastrous and will produce a large number of FA.

### 2.3.2.2  Dataset Design Considering of Irregularly Wearing (IW)

We define irregularly mask-wearing as shown in Figure 2.3 (c), in which people wear a mask but part of their nose or mouth is exposed. These samples are extremely similar to mask-wearing which should be categorized as not wearing mask. If we train model on the dataset taking irregularly mask-wearing as positive sample, some images of correctly mask-wearing will be mistakenly identified as not wearing mask, resulting in FA. Therefore, we extract this kind of data separately and classify it as uncertain category, as shown in Figure 2.2 (b).

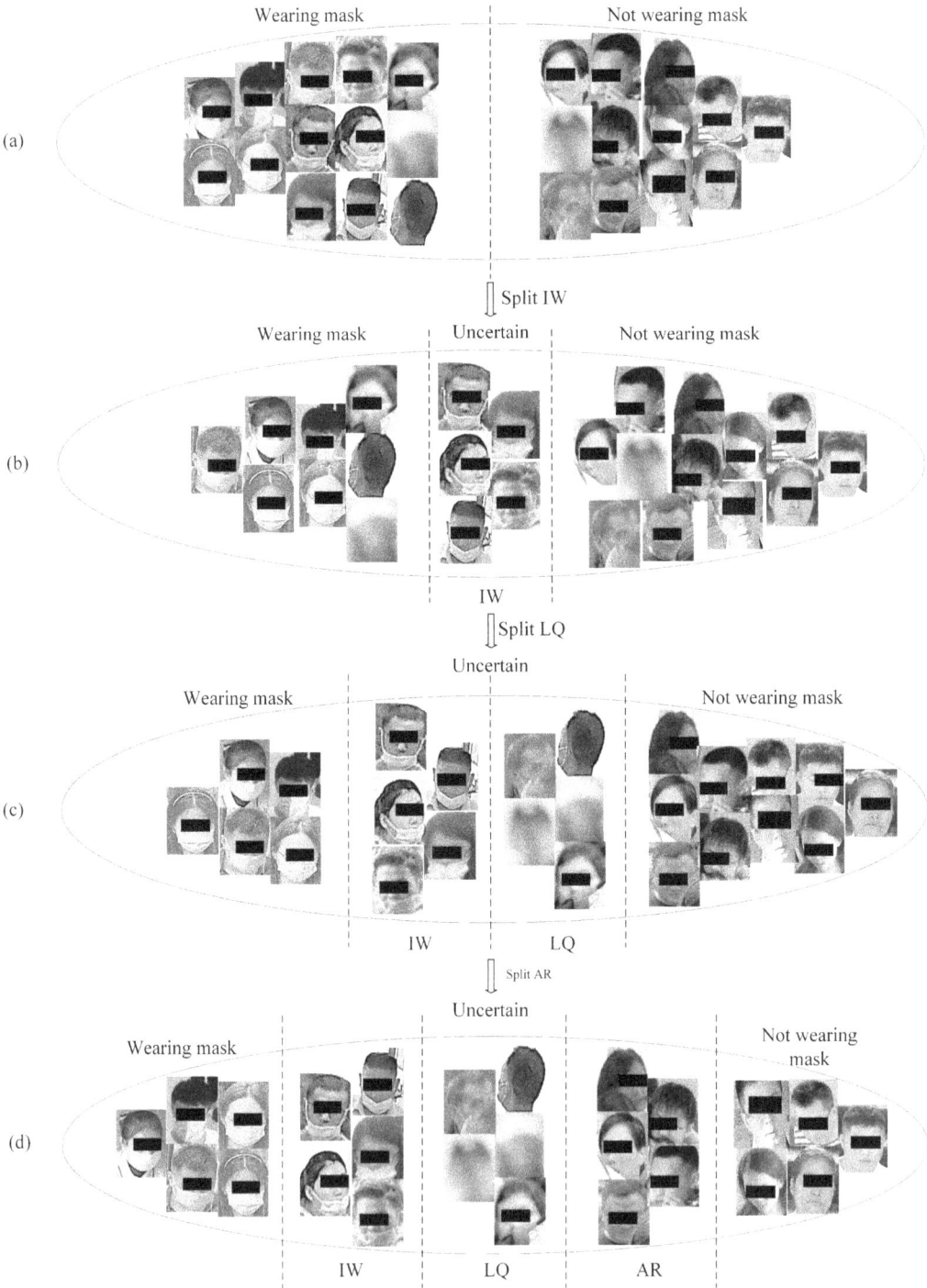

Figure 2.2 The architecture of our framework. Construct uncertain category datasets based on data attributes and application requirements. (a) Straightforward method of dataset design. In (b) we extract the samples with irregular wearing from (a), the low-quality samples are extracted in (c), and we extract the samples with mask-like occlusion in (d).

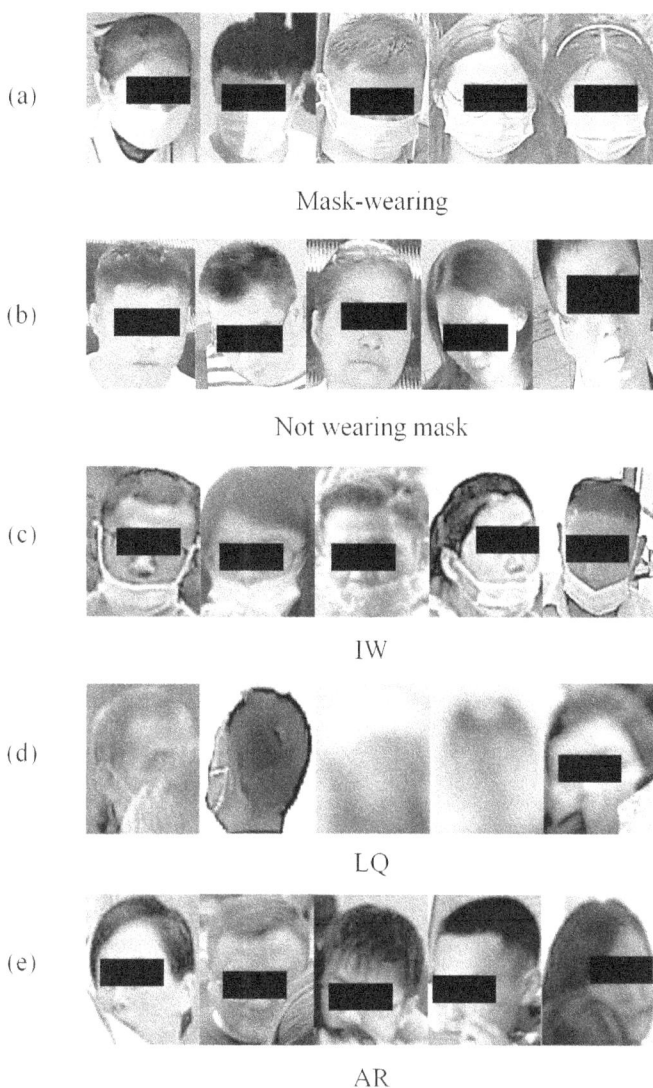

Figure 2.3 Example of all category images. (a) Typical samples of not wearing mask. (b) Representative samples of mask-weaing. (c) Irregularly wearing: people wear a mask but part of their nose or mouth is exposed. (d) Low Quality: occlusion, overexposure, blurring and so on make it impossible to confirm whether a person is wearing a mask or not. (e) Mask-like occlusion: mouth or nose is subtle occluded by non-mask objects.

### 2.3.2.3  Dataset Design Considering of Low Quality (LQ)

Low-quality data means that it is hard to identify whether a person is wearing mask or not due to shooting angle, illumination, distance, occlusion, etc. As shown in Figure 2.3 (d), factors like occlusion, overexposure, blurring, only the back of people's head or part of the side face can be seen and so on make it impossible to confirm whether a person is wearing a mask or not. Therefore, the above data needs to

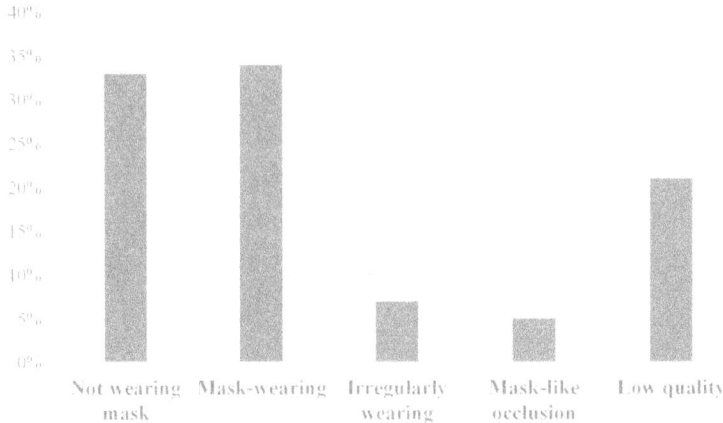

Figure 2.4   Distribution of training dataset.

be dealt with separately and classified as the uncertainty category, as illustrated in Figure 2.2 (c).

### 2.3.2.4   Dataset Design Considering of Mask-Like Occlusion (MLO)

Mask-like occlusion data refers to the mouth or nose being subtle occlusion by non-mask objects as illustrated in Figure 2.3 (e). Mask-like occlusion data is classified as negative samples according to the traditional data design method, which may lead to incorrect recognition of mask-wearing samples similar to MLO. However, extracting MLO data will decrease the recall of not wearing mask. There are some actual requirements that are less stringent for recall rate of not wearing mask yet have high expectations for precision rate, because frequent FA brings very bad user experience. Hence, we need to further separate MLO data out and then put them to the uncertainty category (Figure 2.2 (d)).

## 2.4   EXPERIMENTS

In this section, we conduct experiments on a collected mask-wearing recognition dataset containing more than 10,000 images for training and 1,000 images for testing which do not overlap the training set. We use resnet50 [78] as backbone and softmax loss as loss function. We fine-tuned model on ImageNet's pre-trained model, with training learning rate being 1e-3, cosine decay applied, optimizer Adam [110], and we trained for 50 epochs in total.

According to our design, the training dataset is divided into five parts(as shown in Figure 2.4), of which 3384 are not-wearing-mask, 3465 are mask-wearing, 587 are irregularly wearing, 2319 are low quality, and the remaining 375 are mask-like occlusion. We unify the irregularly wearing, low-quality, and mask-like occlusion into the uncertainty class for training.

We demonstrate the effectiveness of our dataset design method through ablation experiments. The results of our experiments are described in Table 2.1. Specifically,

TABLE 2.1    Results on our designed dataset.

| Dataset Design | FAR |
| --- | --- |
| Original | 8.9% |
| Extract IW only | 4.1% |
| Extract LQ only | **3.8%** |
| Extract MLO only | 5.8% |
| Extract IW+LQ only | 1.4% |
| Extract IW+MLO | 3.6% |
| Extract LQ+MLO | 3.3% |
| Extract IW+LQ+MLO | **0.8%** |

when the uncertainty class contains any one of IW, LQ, and MLO, the FAR (false alarm rate) decreases. This indicates that IW, LQ, and MLO all have an impact on FA, and it can be seen from the table that LQ has the greatest impact on FA. When the uncertainty class contains any two of IW, LQ, and MLO, the FAR decreases more. When the uncertainty class contains IW, LQ, and MLO all, the FAR reaches the minimum value of 0.8%, which indicates that our dataset design method can greatly reduce FA.

## 2.5    CONCLUSION

In this chapter, we design the dataset finely driven by application, based on the data-centric AI paradigm. We take the mask-wearing recognition as an example, and collect and open source mask-wearing recognition dataset with more than 10 thousand images covering various application scenarios. Compared with the mainstream dataset design schemes, our method achieves better results and effectively reduces FA. Additionally this dataset design method can be extended to various scenarios of industrial applications.

# CHAPTER 3

# Hand-Held Action Detection Paradigm for Smoking Detection

## 3.1 INTRODUCTION

In practical applications especially in industrial scenarios, some hand-held human actions need to be monitored closely, including smoking cigarettes, dialing, eating, etc. Under most circumstances, smoking cigarettes is an important problem of safety. Places like chemical plants, industrial workshops and petrol stations, all require strict control of smoking cigarettes, which may bring about fire, explosion, etc. For instance, a spark in some chemical plants may lead to a disaster of a lot of human life. Additionally, in most scenarios, it is not permitted to use smart phone or eating something during on-duty. In this chapter, we take smoking cigarettes as a typical example to discuss how to detect this kind of hand-held action.

With the rapid development of deep learning [194], more and more detection methods are utilized in the field of smoking cigarettes detection [160]. However, most of these methods adopt only one single detection model, which may conquer one kind of problem but meanwhile lead to another problem. In other words, due to the inherent attributes of the target object, methods utilizing one single model are scarcely possible to cover all kinds of problems that may occur under real-world scenarios.

Intuitively, cigarettes smoking detection methods can be roughly divided into two types: one type of method focuses on the cigarettes themselves, while the other focuses on the general smoking pose of humans, as shown in Figure 3.1. Typical distribution of real-world scenarios' smoking cigarettes can be roughly classified into three classes: (a) images having no cigarettes but human body pose are similar to smoking cigarettes; (b) images having cigarettes and are easy to be detected correctly; (c) images having sticks similar to cigarettes but having no cigarettes. As for these different kinds of scenarios, current mainstream methods can hardly cover all these cases and achieve a high accuracy. This is because the class (a) and (c) are exactly two opposite directions for training models: a model focusing its attention on class

DOI: 10.1201/9781003644972-3

Figure 3.1   How to detect smoking cigarettes in real-world scenarios? Generally, one may focus more on the cigarette itself while the other on the human smoking pose.

(a) is natively inclined to neglect the small cigarettes compared to the much bigger human body, and vice versa. For instance, as is shown in Figure 3.2, if one method focuses more attention on the cigarette itself, it's more inclined to falsely take sticks that are similar to cigarettes as the final target; furthermore, as is shown in Figure 3.3, if one method focuses its attention on the overall human body and smoking pose so as to amend the former stick mistakes, it again may bring about other problems: it may overlook the cigarettes and take the bigger pose and human body as final target instead.

To address these issues, the hierarchical coarse-to-fine detection framework is proposed in this chapter as a new application-driven AI paradigm. In this framework, the coarse detection module detects the target of human smoking pose consisting of the whole hand, cigarette and head, while the followed fine detection module

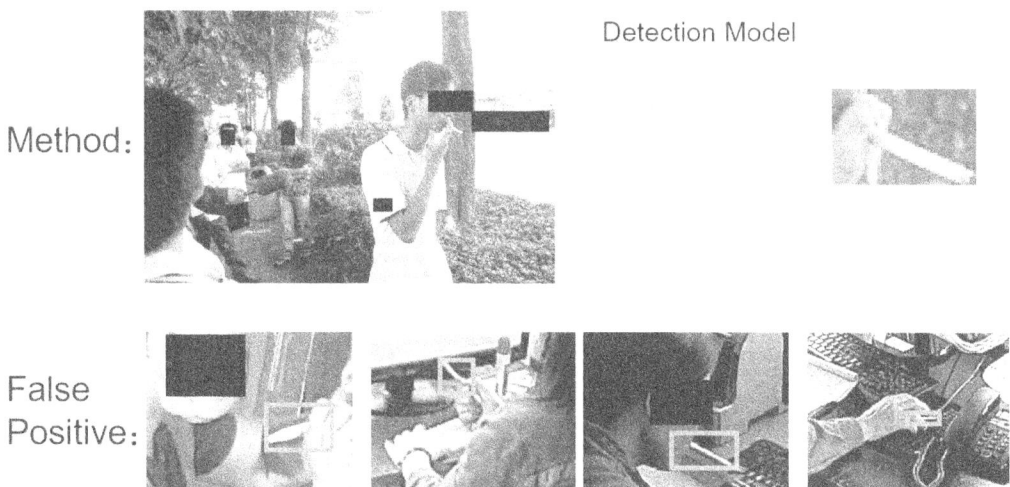

Figure 3.2   Method focusing only on the cigarette itself is more inclined to falsely take sticks that are similar to cigarettes as targets.

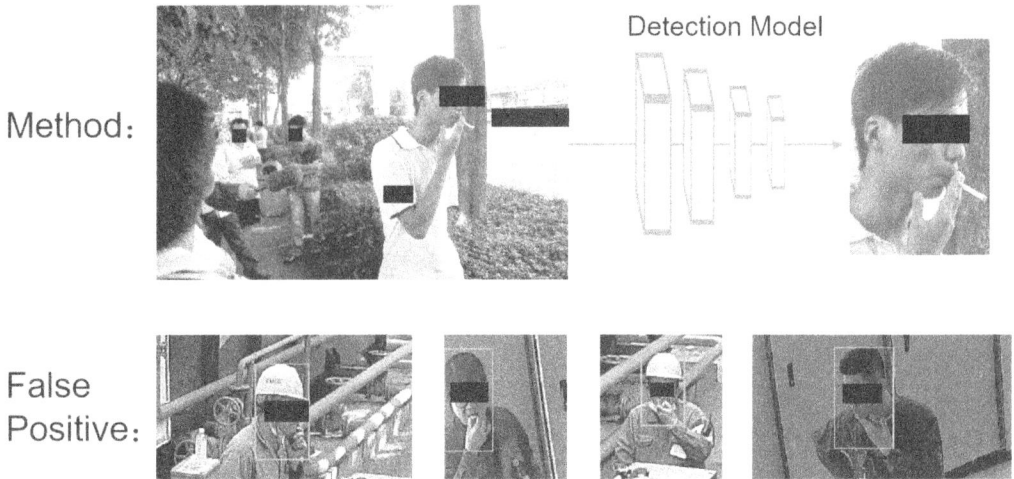

Figure 3.3 Method focusing on the human smoking pose may overlook the cigarettes and take only the body pose as target.

detects the target of cigarette consisting of the fingers holding cigarette, mouth area, and the whole cigarette. With the hierarchical framework, both the overall pose and details are considered, which confirms the high accuracy. The rest of the chapter is arranged as follows. The related work is introduced in Section 2. In Section 3, the proposed application-driven AI paradigm is presented in detail. The experiments are done and results are given in Section 4. Finally, in Section 5, the conclusion is drawn.

## 3.2 RELATED WORK

Some methods have been proposed to handle the problem of detecting smoking cigarettes. The method [250] proposes a smoking image detection model based on a convolutional neural network, referred to as SmokingNet, which automatically detects smoking behaviors in video content through images, this method can detect smoking images by utilizing only the information of human smoking gestures and cigarette image characteristics without requiring the detection of cigarette smoking. The method [199] proposes a novel algorithm for automatic detection of puffs in smoking episodes by using a combination of Respiratory Inductance Plethysmography and Inertial Measurement Unit sensors. The detection of puffs was performed by using a deep network containing convolutional and recurrent neural networks.

## 3.3 THE PROPOSED APPLICATION-DRIVEN AI PARADIGM

### 3.3.1 The Hierarchical Object Detection Framework

In this chapter, we propose a coarse-to-fine two-stage method to deal with this dilemma. Our method utilizes two detection modules: the first module focuses on a coarse object of human smoking pose, while the followed second one focuses on the

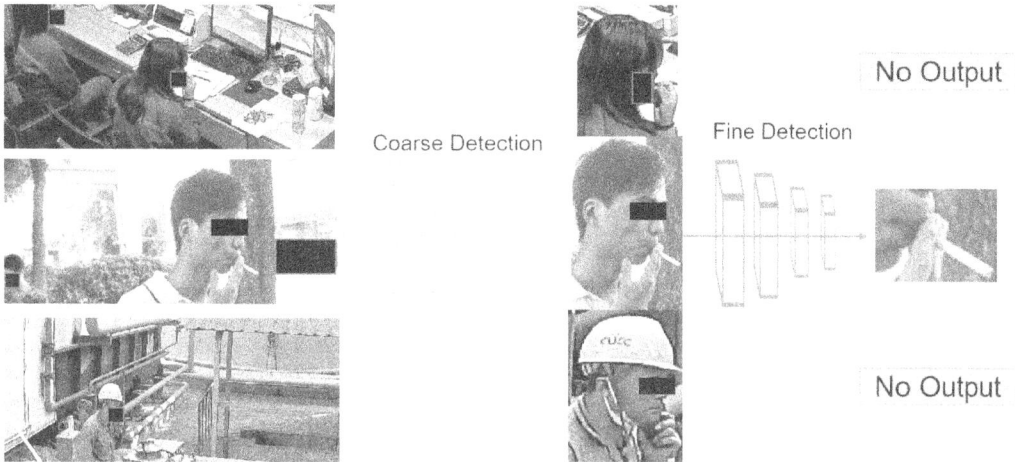

**Figure 3.4**  Our hierarchical object detection framework composed of two detection modules. The first module focuses on a coarse object consisting of the whole hand, cigarette and head. The followed second module detects the object consisting of the fingers holding cigarette, mouth area and the whole cigarette.

fine object of the cigarette. Overview of the two stages is shown as Figure 3.4 and detailed as follows.

### 3.3.2  Coarse Detection Module

The coarse detection module roughly detects an object including whole hand, cigarette, and head, as shown in Figure 3.5. In this detection, we take a typical human pose of smoking cigarettes as target. A pose of smoking consists of several parts, but the typical and essential one is to hold a cigarette and feed it into mouth, which we take as the target of the first stage's detection module. In this way, hands and whole head, as context of a typical pose of smoking cigarettes in an image, can filter out most targets that can be easily misdetected as smoking cigarettes.

### 3.3.3  Fine Detection Module

The second module detects a finer object consisting of the fingers holding cigarette, the mouth area, and the whole cigarette. Following the coarse detection module, the second module detects a much smaller object, as shown in Figure 3.6. In this way, the second module's results, which are based on the first module's input, can natively utilize the first module's design philosophy to filter out a majority of mistakes, meanwhile the well-designed finer object can filter out other kinds of false detections: for instance, due to the relatively bigger target, the first module may improperly take hands and a whole head as the target of smoking cigarettes and ignore the cigarettes. This is because in the bigger target, the cigarette itself is a much smaller object compared to hands and human head, after several CNN layers, the feature map may just take the cigarette as an irrelevant noise of the overall target. Hence the second module concentrates more on a rather small target: the fingers holding cigarette, mouth area

Figure 3.5 Examples of the coarse detection model's annotation. The target object consists of the whole hand, cigarette and head.

and the whole cigarette. In this smaller area, the module will give more attention to the cigarettes, while still give consideration to other body parts of fingers and mouth, which in all are pivotal components of a pose of smoking cigarettes.

### 3.3.4 Detection Model

In our proposed framework, various kinds of object detection models can be adopted either for the coarse detection module or the fine detection module, and it's not limited to one certain object detection model. In this chapter, we utilized the basic object detection model YOLOv5 [101] and faster rcnn [182] to construct the coarse detection model and fine detection model for our ablation study experiments.

## 3.4 EXPERIMENTS

In this section, we will firstly introduce our dataset, which we collected from all kinds of real-world scenarios, including chemical plants, industrial workshops, and petrol stations, etc. Then we will describe the details of our experiments.

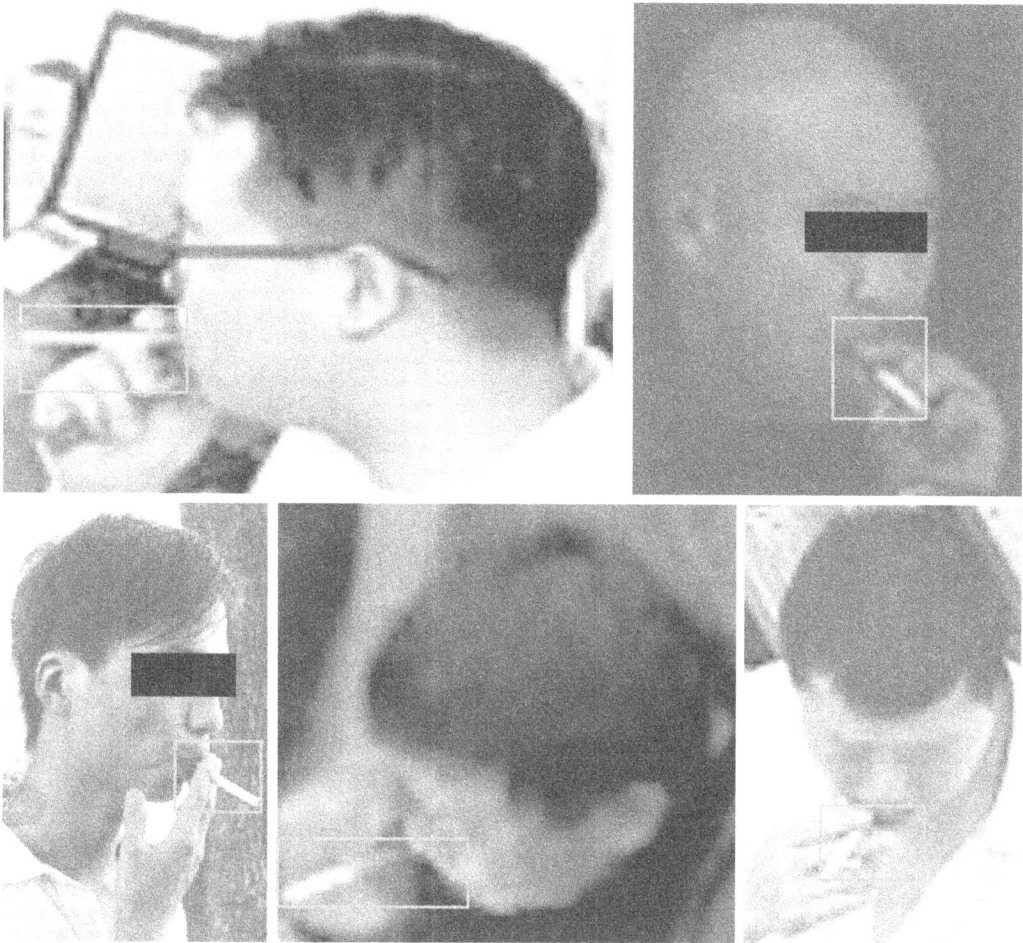

Figure 3.6 Examples of the fine detection model's annotation. The target object consists of the fingers holding cigarette, mouth area and the whole cigarette.

### 3.4.1 Dataset

#### 3.4.1.1 Coarse Detection Model's Dataset

We collected images from public webs, manual simulation, and certain real-world scenarios, including chemical plants, industrial workshops, and petrol stations, etc. The target object consists of the whole hand, cigarette and head. These images cover various gestures, angles, facial features, ages, and illuminations of smoking cigarettes, which can enhance the detection models' generalization ability while simultaneously maintaining a rather high accuracy. The smoking poses in these images are annotated by the way, as shown in Figure 3.5.

#### 3.4.1.2 Fine Detection Model's Dataset

Based on the coarse detection model's dataset, the fine detection model's dataset is constructed by applying the coarse object detection and fine annotation. The target object consists of the fingers holding the cigarette, mouth area, and the whole cigarette. And cigarettes are annotated by the way, as shown in Figure 3.6.

### 3.4.1.3 *Additional Dataset*

Smoking cigarettes, as a safety event which requires intense and precise attention and monitoring, sometimes requires high precision under some circumstances while sometimes requires high recall under other circumstances. In view of this realistic dilemma, we additionally collected images without cigarettes inside but are very likely to be classified as smoking cigarettes. In this way, the dataset can enhance the models' generalization ability to certain degree, which will increase precision and recall simultaneously.

### 3.4.2 Experimental Settings

The coarse detection model's input size is 1280x1280, and batch size is 64, and we train the model on a 8*V100 machine. In view of the fine detection model's target is detected based on the coarse model's output, which most of the time is rather small, we set the fine model's input size as 320x320, and batch size 100, and we train the model on a 4*2080Ti machine. For both models, we use warmup epochs as 2, momentum for warmup epochs is 0.5, learning rate is set as 0.0032, and momentum is 0.843.

### 3.4.3 Experimental Results

In this section, we conduct experiments on our collected dataset. Both the coarse model and fine model are trained using yolov5m and faster rcnn. Additionally, to compare with our coarse-to-fine models, the single model focusing on human smoking pose and the one focusing on cigarettes are trained using yolov5m and faster rcnn.

After training the models, we conduct ablation study on the models to show the superiority of our framework. We manually choose 450 positive images containing positive targets, and 400 negative images which contain no positive targets but are very likely to be detected as smoking cigarettes.

The results of YOLOv5 and faster rcnn are shown as Table 3.1, Single Model I focuses on the overall human smoking pose, while Single Model II focuses more

TABLE 3.1   Accuracy on test data for different frameworks. Single Model I focuses on the overall human smoking pose, Single Model II focuses more on the cigarette itself, and Coarse-to-fine Models is our proposed framework.

| Models | Frameworks | Accuracy |
|---|---|---|
| | Single Model I | 0.714 |
| Yolov5 | Single Model II | 0.753 |
| | **Coarse-to-fine Models** | **0.921** |
| | Single Model I | 0.704 |
| Faster RCNN | Single Model II | 0.734 |
| | **Coarse-to-fine Models** | **0.913** |

False Positive of
Single Model I:

False Positive of
Single Model II:

Figure 3.7  Examples of images containing false positive target. Single model I and single model II both make mistakes as a result of its intrinsic shortages.

on the cigarette itself. The results of both YOLOv5 and faster rcnn have the same regular pattern. In detail, for the 450 positive images, Single Model I, Single Model II, and our Coarse-to-fine Models both achieve satisfactory results; as for the other 400 negative images; however, the two single models both detect many targets as smoking cigarettes, while the images have no smoke at all, which makes much more mistakes than our coarse-to-fine model.

In Figure 3.7, we show several typical mistakes of detecting images containing false positive targets as smoking cigarettes. For instance, take the top left image in Figure 3.7 as an example, Single Model I detects the target as a true positive target, in which the person inside feeds nothing into the mouth but the action pose is similar to smoking cigarettes; however, in our method, the fine model correctly ignores this target and detects nothing as a true positive target, in other words, the person with the smoking pose but without cigarette is detected as a true negative target, which is classified correctly. In this way, a method utilizing one single model focusing on overall human smoking pose makes a mistake while our method does it precisely correctly.

Furthermore, take the top right image in Figure 3.7 as an example. Single Model II detects the target as a true positive target, in which the person inside is holding a pen only; the action pose is not similar to smoking cigarettes, but the pen itself looks like a cigarette; however, in our method, the coarse model, which focuses on the overall human smoking pose, correctly ignores this target and detects nothing as a true positive target. In this way, a method utilizing one single model focusing on the cigarette itself makes a mistake, while our method does it precisely correctly.

## 3.5 CONCLUSION

In this chapter, we propose a hierarchical object detection framework for hand-held action detection, which is composed of a coarse detection model to localize the human action pose and the fine detection model to identify the object itself. Taking smoking cigarette detection for example, the dataset is collected from various practical application scenarios and annotated in both a coarse manner and a fine manner. Based on typical basic models YOLOv5 and faster rcnn, the coarse model and fine model are trained, and compared with single-model frameworks. The experimental results show that the coarse-to-fine framework achieves better results and effectively reduces false alarms compared to the single-model frameworks. Our framework, as a new application-driven AI paradigm, can be further generalized to handle various kinds of hand-held action detection such as smoking, dialing, and eating.

# CHAPTER 4

# Human Action Recognition Paradigm for Person Safety Supervision in Various Practical Scenarios

## 4.1 INTRODUCTION

In practical applications, various scenarios require the monitoring of human actions. In contexts involving safety supervision, surveillance cameras capture image data from the work areas of safety supervisors to ensure their presence on duty and to ascertain whether they are engaged in their responsibilities or potentially sleeping. Additionally, certain situations require limitations on personnel numbers; for instance, it is essential to monitor worker statistics within factory assembly line environments to maintain compliance with established personnel limits. In chemical plant settings, fall detection systems can promptly identify abnormal human postures, thereby providing timely alerts regarding potential gas leakage incidents. Consequently, human action recognition holds significant research importance for mitigating safety accidents across a range of scenarios.

There are many studies on human action recognition. Two-stream RNN/LSTM framework [203, 104, 12] takes different input features extracted from the RGB videos for Human Action Recognition (HAR) and gets human action results through fusion strategies. However, the computational complexity of the two-stream framework is huge compared to a single CNN framework, and it relies on continuous image input. 3D CNN framework [94, 251, 218] extends 2D CNNs to 3D structures, to simultaneously model the spatial and temporal context information in videos. However, this algorithm is also computationally intensive and relies on continuous image input. Skeleton-based methods [201, 68, 136, 208] extract skeleton sequences to encode the trajectories of human body joints, which characterize informative human motions. However, these algorithms are not only computationally intensive but also unstable in actual monitoring scenarios. Numerous works [63, 139, 126] have made great

DOI: 10.1201/9781003644972-4

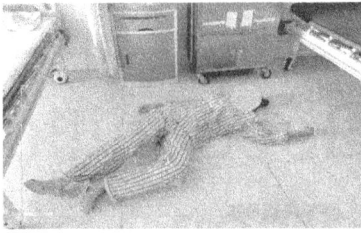

(a) Fall detection     (b) Sleep detection

(c) On-duty detection

Figure 4.1 Human action recognition in different scenarios depends on different visual information including the whole body, upper body, or part of body.

contributions to the improvement of dl-based object detection algorithms. Regression-based algorithm [226] directly estimates the count of pedestrians. The method [8] detects fall actions via SVM. And skeleton-based and human key-points-based algorithms [252, 235] are proposed for fall detection.

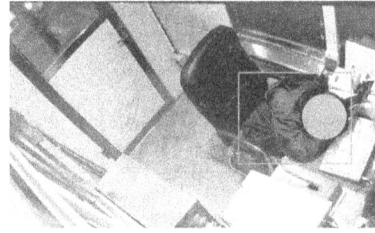

Generally, human action recognition heavily depends on captured visual information. As shown in Figure 4.1, some scenes require the whole body to make a decision (e.g., fall detection), while others rely on the upper body (e.g., sleeping detection), or even just part of the body (e.g., person counting for on-duty detection). Currently, most algorithms handle each type of scene independently; however, a unified framework is desirable to support multiple scenarios simultaneously. Additionally, inspired by the fact that humans can often recognize actions from a single instance of visual information, accurate identification using a single video frame may be feasible, potentially reducing the computational cost of processing temporal sequences. Numerous works [63, 139, 126] have significantly improved deep learning-based object detection algorithms by enhancing both accuracy and processing speed. Regression-based algorithms [226] directly estimate pedestrian counts by learning a mapping from image features to the number of people present. The method [8] detects fall actions using SVM. Additionally, skeleton-based and human key-point-based algorithms [252, 235] have been proposed for fall detection.

To address the above-mentioned issues, we propose a unified human action recognition framework composed of a multi-form human detection module and an action classification module. With a single image as input, the multi-form human detection module simultaneously localizes the whole body, upper body, and parts of body with different labels. The localized human regions are then passed to action classification modules, which identify actions within the image. This scheme processes input images frame-by-frame and assigns action labels to each frame. Combining the results for a time period, the human actions are predicted. The main contributions of this chapter are as follows: 1) A unified framework for human action recognition is proposed as a new application-driven AI paradigm; 2) A multi-form human detection model is introduced, and the corresponding dataset is open-sourced; 3) Experiments demonstrate the effectiveness of the proposed paradigm.

The rest of the chapter is arranged as follows. In Section 4.2 reviews related works. Section 4.3 provides a detailed description of the proposed AI paradigm. Section 4.4 presents the constructed multi-form human detection dataset in detail. Section 4.5 presents the experimental results that demonstrate the effectiveness of the proposed AI paradigm. Finally, Section 4.6 provides the conclusions.

## 4.2 RELATED WORK

Human Action Recognition (HAR), which involves recognizing and understanding human actions, is essential for various real-world applications. HAR can be applied in visual surveillance systems [134] to detect hazardous human activities and in autonomous navigation systems [146] to monitor human behaviors, ensuring safe operation. Additionally, HAR is important for various other applications, such as video retrieval.

Initially, most studies focused on using RGB or grayscale videos as input for HAR [170], owing to their widespread availability and ease of access. In recent years, there has been an emergence of studies [46, 137, 177] that use other data modalities, such as skeleton data, infrared sequences, point clouds, event streams, audio, acceleration, and radar. This is primarily attributed to the development of various advanced and cost-effective sensors.

Multi-frame dense optical flow is primarily used to train a two-stream CNN [201], where the temporal stream handles motion in the form of dense optical flow, and the spatial stream processes still video frames. The two-stream 2D CNN framework [203, 104, 12] generally includes two branches, each taking different input features extracted from RGB videos for HAR. The final result is typically obtained through fusion strategies. Additionally, RNNs are employed to analyze temporal data because of the recurrent nature of their hidden layers. Most existing methods adopt gated RNN architectures, such as Long Short-Term Memory (LSTM) [50, 209], to model long-term dependencies in video sequences.

Numerous studies [94, 251, 218] have extended 2D CNNs to 3D structures to jointly model spatial and temporal context in videos, which is crucial for HAR. The Transformer [219] is a novel deep learning model that has recently emerged as a leader in the machine learning field. The transformer consists of an encoder and a decoder. This architecture allows the transformer to excel in long-term dependency modeling, multi-modal fusion, and multi-task processing [107, 74].

Skeleton-based algorithms encode the trajectories of human body joints, which represent key human motions. Therefore, skeleton data is an effective modality for HAR. Skeleton data can be obtained using pose estimation algorithms on RGB videos

[68, 208]. Most recent works on skeleton-based HAR have used skeleton data obtained from RGB videos [239].

Human detection algorithms are typically classified into one-stage [139, 179, 180] and two-stage object detectors [63, 64, 182]. One-stage detectors are primarily categorized into two types: anchor-based [223, 189, 145, 212] and anchor-free [51, 115, 116, 252, 262]. Currently, the most accurate real-time one-stage human detectors are the anchor-based EfficientDet [212], YOLOv4 [223], and PP-YOLO [145].

Early approaches for person-counting rely on detections [126, 135, 172]. The regression-based approach directly estimates the pedestrian count from the image [226, 22, 23], which limits its scalability for localization tasks. Recently, person counting via density map estimation has emerged as a promising approach, where the input image is processed into a crowd density map, which is then integrated to estimate the number of people in the image [9, 19, 167, 260].

Early methods for sleep and fall detection rely on SVM [102, 8], KNN [73, 41], and feature-threshold approaches [237, 33]. These methods use foreground extraction through background subtraction. CNN-based approaches [127, 56, 178] extract features using convolutional networks. Skeleton-based or keypoint-based approaches are also popular [252, 235]. They extract features from skeletal points' positions, angles, and scales to characterize human actions.

For efficient HAR in various scenarios, the method in Figure 4.2 facilitates straightforward action recognition, where human detection is applied to an input image followed by action recognition. Intuitively, different actions require different body information to recognize the action. For example, whole-body detection is required for fall detection, upper-body detection for sleep detection, and part-body

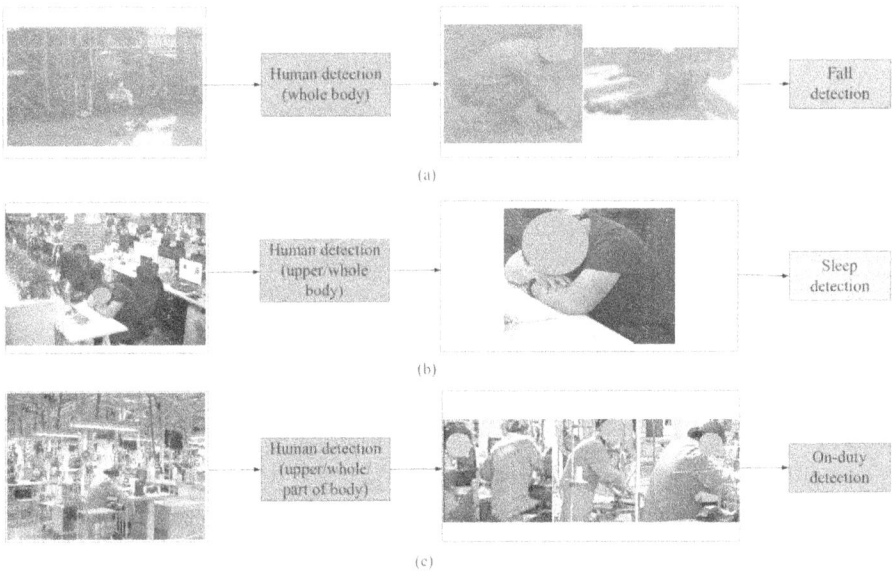

Figure 4.2 The straightforward human action recognition scheme with different human detection for different action scenarios: (a) fall detection, (b) sleep detection, and (c) on-duty detection.

detection for on-duty tasks. Thus, different human detection setups are needed for different actions. In this chapter, we propose a general human detection method to support various actions, which is then used to construct a unified action recognition framework. Additionally, the method is extended from a single image input to a video sequence in a computationally efficient manner.

## 4.3 THE PROPOSED APPLICATION-DRIVEN AI PARADIGM

### 4.3.1 System Overview

The proposed application-driven AI paradigm for HAR, as shown in Figure 4.3, consists of two stages: multi-form human detection and action classification. In the first stage, human body regions are identified from an input image with three types of labels. In the second stage, the corresponding human body regions are used to classify actions using an action classifier. Thus, for each input image, the system outputs the confidence score or action label for each action type being considered. For the video sequence comprising temporal image frames, the global action label is determined by finding the most frequent action label. Instead of processing the entire video, periodic frames are used. In most cases, a single frame suffices for action recognition in the video. In this work, we use the minimum number of frames, reducing computational time.

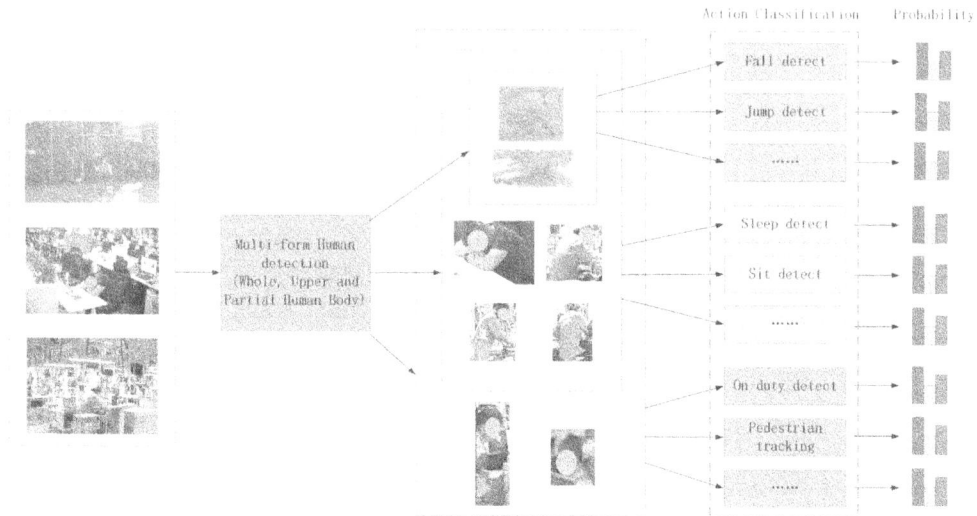

Figure 4.3 The application-driven AI paradigm for human action recognition. With images as input, the multi-form human detection model produces three kinds of human body regions. Then, various action classifiers are followed corresponding to certain kinds of human body regions.

### 4.3.2 Multi-form Human Detection

In the proposed multi-form human detection, three kinds of human body regions, i.e., the whole body, upper body, and part of body, are localized from an input image. These human body forms may cover nearly all potential application scenarios. For example, cases such as on-duty detection or pedestrian tracking requires only part of body to make a decision, while such case as sleep detection or sit detection often requires the upper body for decision-making. Furthermore, cases such as fall detection, jump detection, running detection, or standing detection typically require the whole body for decision-making.

In general, an object detection model is applied for multi-form human detection. Here, we employ YOLOv5 [223], an object detection algorithm that divides an image into a grid system, where each cell in the grid is responsible for detecting objects within its area. As the backbone, we use the new CSP-Darknet53 module [224], SPPF [77], and New CSP-PAN [138] as the neck module. Considering the trade-off between speed, memory consumption, and accuracy, we use the m version of the YOLOv5 model with a resolution of 640x640, which achieves 20fps in a non-GPU environment, such as with an Intel i7-1185g7 processor. By using the iGPU of Intel CPU, the speed of our algorithm will be increased by 50% to reach 30fps. To train the object detection model, we construct a new dataset, which will be presented in detail in Section 4.4.

### 4.3.3 Action Classification

In action classification, the action classifier determines whether each detected human body region corresponds to a specific action. Various action recognition scenarios can be supported if the appropriate action classifiers are provided. It is important to note that different action scenarios may require different types of human body regions. For example, fall detection requires the whole-body region, sleep detection requires both the upper and whole-body regions, whereas on-duty detection requires either the part-body, upper-body, or whole-body regions.

Generally, a deep learning-based image classification model can be used to design the action classifier. However, some important considerations should be carefully addressed when preparing the training dataset. For example, in the case of sleep detection, we need to extend the two classes of sleep/nonsleep to three classes of sleep/sit/nonsleep or more, because the posture of a person sitting upright may resemble the posture of a person lying on a table sleeping. In the case of fall detection, the posture of a person sitting on a bed/chair/sofa may resemble a fall, so we need to increase the number of classes in the classifier to minimize false positives. In this work, we use ResNet18 [78] as the classification model, considering the trade-off between accuracy and speed. In contrast, for on-duty detection, person counting based on person detection may be sufficient to determine whether the number of persons is reasonable at work.

TABLE 4.1   Human annotation statistics.

|  | Whole body | Upper body | Part of body |
| --- | --- | --- | --- |
| Quantity | 102000 | 68000 | 105000 |

## 4.4   OPEN-SOURCE OF MULTI-FORM HUMAN DETECTION DATASET

To train the proposed multi-form human detection model, we construct an open-source dataset: we collect and annotate 700 videos from actual applications containing human bodies with different poses in various scenarios to improve the algorithm's performance, then we distinguish three kinds of human body labels, i.e., whole body, upper body, and part body. In total, we have collected 35,000 images containing 275,000 human bodies. This dataset covers whole, upper, and part bodies with various angles, illuminations, sizes, and qualities, and is manually labeled by several people in numerous rounds. As shown in Table 4.1, there are a total of 275,000 annotations: 102,000 for whole bodies, 68,000 for upper bodies, and 105,000 for part bodies.

### 4.4.1   Annotation for Whole Human Bodies

For the whole body annotation, we define the strategy that the torso and legs are mostly not occluded. Specifically, unlike the upper body annotation in Section 4.4.2, the labeling of the whole body focuses more on the torso and lower limbs. Even if the head and hands are occluded, when the torso and lower limbs of the person are visible, we still annotate it as the whole body. Finally, we obtained 102,000 whole body annotations. Some samples are shown in Figure 4.4, where the whole bodies are annotated with boxes.

Figure 4.4   Examples of whole human body annotations. Among these pictures, three whole human bodies are annotated with boxes.

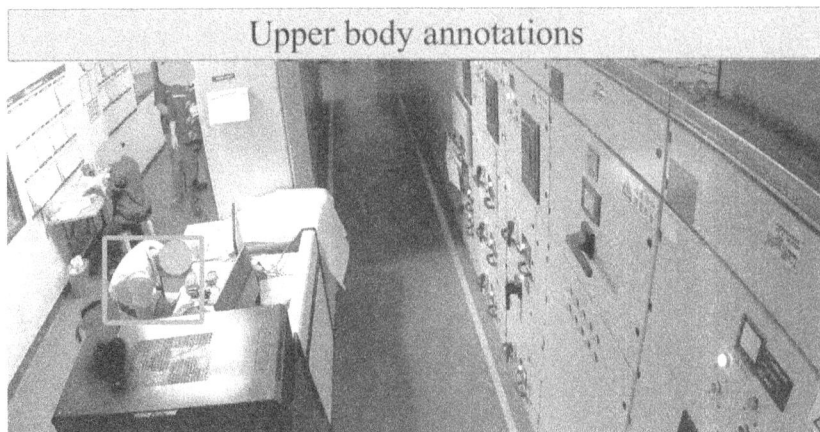

Figure 4.5   Examples of upper human body annotations. Among these pictures, one upper human body is annotated with box.

### 4.4.2   Annotation for Upper Human Bodies

For the upper body annotation, we set the strategy that at least the head and shoulders must be visible. This annotation strategy to the training data finally leaves us with 68,000 upper body annotations. Figure 4.5 shows some examples of upper body annotations, highlighted with boxes.

### 4.4.3   Annotation for Partial Human Bodies

For the partial body annotations, human bodies that are not labeled in Sections 4.4.1 and 4.4.2 are labeled. Finally, we obtained a total of 105,000 partial human body annotations. Some examples are shown in Figure 4.6, with the partial bodies annotated with boxes.

### 4.4.4   Annotation for Action Classification

Using the annotated human bodies, we create action subsets for training classifiers. The five subsets include: standing, jumping, falling, sleeping, and sitting. These subsets can be used to train action classifiers.

## 4.5   EXPERIMENTS

### 4.5.1   Implementation Details

In the YOLOv5 model for multi-form human body detection, we disable mosaic to avoid incomplete human body labels. We still apply copy-paste and random affine transformations (rotation, scale, translation, and shear). For the rotation parameter, we expand the default range of random sampling from $(-5°, 5°)$ to $(-90°, 90°)$, which significantly improves the recall rate for inclined human bodies. For the multi-scale parameter, which controls image scaling during training, we set it to $(1 \sim 1.5\times)$.

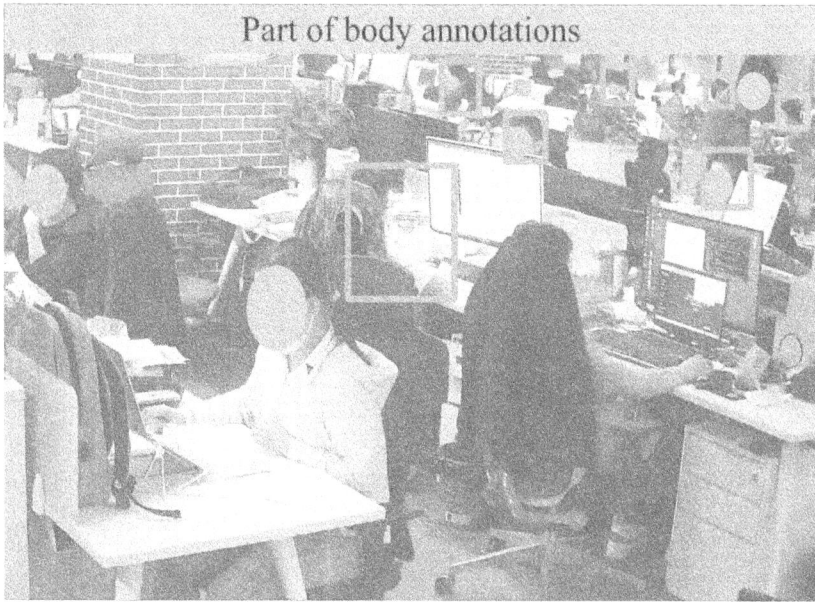

Figure 4.6 Examples of partial human body annotations. Among these pictures, seven partial bodies are annotated with boxes.

For action classification with the ResNet18 model, we modify the input size from $224 \times 224$ to $384 \times 128$, which is more effective when the human body is used as input. During training, we dynamically balance the sample count across categories in each mini-batch.

### 4.5.2 Annotation for Action Classification

Dataset and Metric: We evaluate our method on the public UR Fall Detection Dataset (UR) [114] and the Fall Detection Dataset (FDD) [3]. The UR Fall Detection Dataset consists of 70 sequences, including 30 fall instances and 40 activities of daily living. The dataset contains a total of 22,636 images, with 16,794 images for training, 3,299 for validation, and 2,543 for testing. We use the original partitioning for training our network. We present quantitative comparison results using sensitivity (true positive rate) and specificity (true negative rate) as metrics.

Taking fall detection as an example, we compare the original detection-classification method and state-of-the-art methods [127, 235, 209] with our method, which uses our proposed paradigm. In the original detection-classification method, the YOLO model is used for human body detection, which fails to distinguish between different human body regions for action classification. As shown in Table 4.2, compared to the original detection-classification process, our paradigm significantly improves both specificity and sensitivity. Combining these two metrics, our paradigm achieves the best performance compared to other algorithms. Regarding specificity, it is intuitive that the whole body is more suitable for fall classification than the upper or partial body.

TABLE 4.2   Experiments on UR and FDD datasets.

| Methods | Sensitivity | Specificity |
|---|---|---|
| Original YOLO + classification | 97.9% | 85.3% |
| SVM [127] | 91.1% | 87.3% |
| CNN [209] | 92.8% | 96.1% |
| CNN + human key points [235] | **99.5%** | 98.7% |
| Our method | 99.4% | **99.2%** |

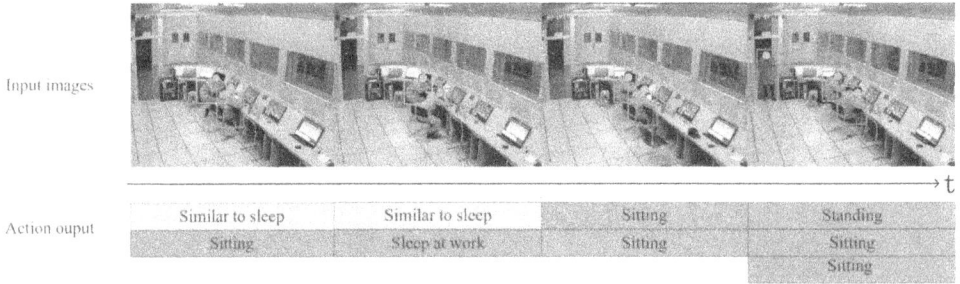

Figure 4.7   Examples of human action recognition in a working environment. In addition to the detection of workers, the system also identifies the action of sleeping at work, and even classifies non-sleeping actions such as standing, sitting, and one similar to sleeping.

### 4.5.3   Experiments on Our Dataset

We collect video data from actual scenarios for action analysis, which forms our dataset. The video data spans 13 hours, consisting of 200 video segments. For training our model, two-thirds of the dataset is used for training, and the remaining third for testing. Figure 4.7 shows an instance of work status surveillance, which can recognize four human actions: standing, sitting, sleeping, and sleeping at work.

We evaluate our detector using the original YOLO detection. In Table 4.3, we show the quantitative comparison results using precision and recall as metrics. Compared to the original detection, we experience a slightly lower recall rate but a significant boost in precision. This shows that detecting the human body in three forms yields a higher precision rate than detecting in only one form.

We evaluate our algorithms with the original detection-classification method. In Table 4.4, we also show the quantitative comparison results of four different action classifications using precision and recall as metrics. Compared to the original detection-classification, we observe a slight decrease in recall rate for three applications and a slight increase in recall rate for sleep detection. We improve the precision

TABLE 4.3   Detectors' experiments on our dataset.

| Methods | Precision | Recall |
|---|---|---|
| YOLO original | 95.9% | **92.2%** |
| Multi-form human detector | **97.1%** | 91.6% |

TABLE 4.4  Classifiers' experiments on our dataset.

| Methods | applications | | | | | | | | | |
| | Fall detection | | Sleep detection | | Jump detection | | On-duty detection | |
| | Precision | Recall | Precision | Recall | Precision | Recall | Precision | Recall |
| Original YOLO + classification | 83.4% | **99.2%** | 77.5% | 96.1% | 95.5% | **98.7%** | 95.9% | **94.5%** |
| Mutli-form human detector + classification (our method) | **97.5%** | 99.0% | **93.1%** | **96.9%** | **98.1%** | 97.9% | **98.8%** | 94.0% |

rate for all applications, particularly for challenging tasks like fall detection and sleep detection, with significant improvements. As shown, action classification based on the three forms of human bodies produced by multi-form human body detection achieves a higher precision rate than classification based on a single form of human body produced by general person detection methods like YOLO.

## 4.6  CONCLUSION

We propose an application-driven AI paradigm for HAR, combining multi-form human detection and action classification. The multi-form human detection model, trained with an open-source dataset, simultaneously localizes three forms of human bodies: whole body, upper body, and partial body. Using the localized human bodies, various action classifiers can be applied to identify corresponding actions. The new open-source dataset is constructed by combining existing open-source datasets and our own collected data, annotated into three forms of human bodies covering nearly all potential applications. Experiment results on both the public dataset and our custom dataset confirm that our system holds significant value in real-world applications. Furthermore, the proposed paradigm has the potential for low-cost implementation, which will be further explored in future work.

# CHAPTER 5

# Person-Counting Paradigm for Intelligent Video Surveillance in Various Practical Scenarios

## 5.1  INTRODUCTION

Person counting is a common application in video surveillance tasks such as visitor analysis, traffic monitoring, and abnormality recognition. In general, the two mainstream techniques for person counting are detection-based methods and density-based methods. The former detects the whole or part of a person's body and counts the number of detected bounding boxes as the final prediction. At the same time, the latter generates a density map and sums up all the pixel values to produce the person-counting estimation.

Over the past few years, many algorithms based on convolutional neural networks (CNN) have been developed to handle various real-world scenarios. For most mid-range scenarios where people are captured from a side view, widely-used object detection methods such as YOLO [179, 180, 181, 13] and SSD [139], pre-trained on MSCOCO [132], already satisfy the requirements. However, in some challenging scenarios, e.g., long-shot images with tiny person targets, top-view images where only heads are visible, or specific places where people are wearing customized suits, additional datasets are necessary for fine-tuning the detector. Furthermore, when facing the scenario of a crowd, the accuracy of almost all person detectors drops significantly due to severe body occlusion that impedes feature extraction of individuals. Compared to detection-based approaches, density-based methods [121, 260, 128, 147, 222, 206] perform better under crowd scenarios.

Although solutions can be found on a case-by-case basis, experienced engineers must review all video streams in advance and manually assign the most suitable person-counting model to each camera device in practical applications. This process is time-consuming and cumbersome, especially as the deployment scale expands.

DOI: 10.1201/9781003644972-5

Moreover, when the backstage manager controls the camera by moving, rotating, or zooming, it can cause a shift in the captured scene, potentially resulting in imprecise output from the predetermined model.

In this chapter, we propose a unified framework that concatenates a scenario classifier and a person-counting module. The scenario classifier is a ResNet-50 [78] network, and the person-counting module contains five fine-tuned models prepared for corresponding scenarios: (i) side-view, where people are captured from a side view in mid-range, (ii) long-shot, where person bodies appear tiny, (iii) top-view, where people are captured from an overhead camera, (iv) customized, where people are wearing specific suits (we use protective suits as an example, which are commonly observed in healthcare institutions), and (v) crowd, where numerous people appear in the image. For (i) to (iv), we utilized YOLOv5, an improved version of YOLOv4 [13], as the person-counting model. For (v), we employ DM-Count [222] as the person-counting model. Additionally, we enhance the robustness of the aforementioned models in specific scenarios by introducing five augmentation datasets. These datasets are also integrated to form a scenario classification dataset used to train the scenario classifier. Further details about these datasets are provided in Section 5.4.1.

The main contributions of this work are three-fold:

- We propose an application-driven paradigm that automatically selects the proper person-counting model for the input image based on the scenario classification result.

- We introduce five augmentation datasets to enhance models in specific scenarios, together with a classification dataset to train the scenario classifier.

- We conduct comparative experiments to demonstrate the generalization of our proposed framework.

The rest of the chapter is structured as follows: Section 5.2 reviews the related work on person detection and crowd counting. Section 5.3 presents the proposed framework and datasets. In Section 5.4, we describe the experiments and demonstrate the results. Finally in Section 5.5, we draw a conclusion.

## 5.2 RELATED WORK

### 5.2.1 Person Detection

Traditional person detection methods relied on handcrafted features, e.g., deformable part models (DPMs) using histograms of oriented gradients (HOG) features [58, 161] and decision forests models using integral channel features (ICF) [45, 254, 256]. With the great success of AlexNet [112] in ImageNet [42] competition, CNN-based methods became mainstream in computer vision tasks like image classification and object detection.

R-CNN [65] pioneered to introduction of CNN into object detection by implementing the two-stage architecture, i.e., a proposal stage and a downstream classification stage. Regarding the intolerable computational complexity for R-CNN to classify each proposal separately, Fast R-CNN [63] optimized the inference time by executing CNN on the whole image at the beginning to extract features shared by all proposals. Faster R-CNN [63] replaced the external proposal modules in Fast R-CNN with a region proposal network (RPN), integrating the two stages into an end-to-end framework that could be trained jointly.

One-stage detection methods completely abandoned the proposal stage. YOLO [179] divided the final feature map into $S \times S$ grid cells and predicted the center coordinates, width, and height of the object bounding box in each cell. However, this approach had a disadvantage in detecting small objects and clustered objects within a single cell. This issue was resolved in YOLOv2 [180] by adopting anchor boxes, whose scales and aspect ratios were set a priori by performing k-means clustering [143] on the training dataset. YOLOv3 [181] introduced residual connections, which was first presented in ResNet [78], to construct a deeper darknet-53 network, and proposed feature pyramid networks (FPNs) for higher recall on smaller objects. YOLOv4 [13] embedded advanced techniques in the backbone, neck, and head of the original YOLOv3 network, and deployed a bag of freebies including mosaic data augmentation, self-adversarial training, and CIoU loss function, together with bag of specials including Mish activation function, cross mini-Batch normalization and drop-block regularization. Just a few weeks later, Jocher et al. released YOLOv5 implemented on PyTorch [165] aiming to further improve accessibility and achieve a greater balance between effectiveness and efficiency. Similar to YOLO serials, SSD [139] also used anchor boxes with a variety of aspect ratios. The difference is that SSD applied anchor boxes on multiple feature maps with different resolutions, thus being able to detect objects of diverse scales.

Person detection can be achieved by training the aforementioned models on datasets taking 'person' as one of the categories, such as MSCOCO [132], Pascal VOC [53] and CityPersons [255]. In these public datasets, the majority of persons were captured from the frontal or side view with an optimal shot range. In recent years, overhead person detection gained significant attention, and models and datasets specifically designed for detecting persons from a top-view perspective were introduced in several studies [4, 171]. However, there was a lack of open-source datasets customized to detect persons in long-range fields and those wearing protective suits, which motivated us to collect and utilize such datasets in our experiments.

## 5.2.2 Crowd Counting

Early approaches to crowd counting relied on detecting persons, heads, or upper bodies [131, 126, 61]. However, these detection-based methods suffered from severe occlusions, especially in dense crowds. Later, researchers developed regression-based frameworks [22, 26, 25, 226] to avoid the detection shortcomings. However, the model was hard to converge without the supervision of head localization annotations in the training process. Recently, methods based on density map estimation [121, 260, 128, 147, 222, 206] outperformed the aforementioned detection-based and regression-based approaches and became the mainstream solution for the crowd-counting problem.

Due to the difficulty of delineating the spatial extent for each person in crowd scenes, existing crowd-counting datasets [260, 88, 89, 230] only mark each person with a single dot on the head or forehead. Consequently, the ground truth density map generated from the annotations is a sparse binary matrix, while the predicted density map is a dense real-value matrix. However, directly measuring the discrepancy between the sparse binary and dense real-value matrices with a loss function can make the network hard to converge.

Therefore, a crucial challenge for all density map estimation methods is to effectively utilize the dot annotations. One common approach is to convert each annotated dot into a Gaussian blob, creating a 'pseudo ground truth' that is more balanced. Most prior methods, e.g., DensityCount [121], MCNN [260] and CSRNet [128] adopted this idea. However, the kernel widths used for the Gaussian blobs may not accurately reflect the size of people's heads in the image, which can significantly impact the network's performance. Another approach is to design a reasonable loss function. For example, a Bayesian loss was proposed in [147] to transform the annotation map into $N$ smoothed density maps, where each pixel value is the posterior probability of the corresponding annotation dot. Recently, the DMCount [222] model used optimal transport (OT) and total variation (TV) loss to measure the similarity between the normalized predicted density map and the normalized ground truth density map. Without introducing Gaussian smoothing operations, DMCount has been shown to outperform the aforementioned Gaussian-based methods.

## 5.3 METHODOLOGY

### 5.3.1 Overview of the Paradigm

The architecture of the proposed paradigm is illustrated in Figure 5.1, and the examples of five scenarios are illustrated in Figure 5.2. Initially, the input image is processed by a scenario classifier, which categorizes the image into one of five predefined scenario types as outlined in Section 5.1. Subsequently, the image is passed to the person-counting module, which comprises five models fine-tuned on scenario-specific augmented datasets. The module automatically selects the appropriate model based on the scenario label. Specifically, for scenarios (i) to (iv), fine-tuned YOLOv5

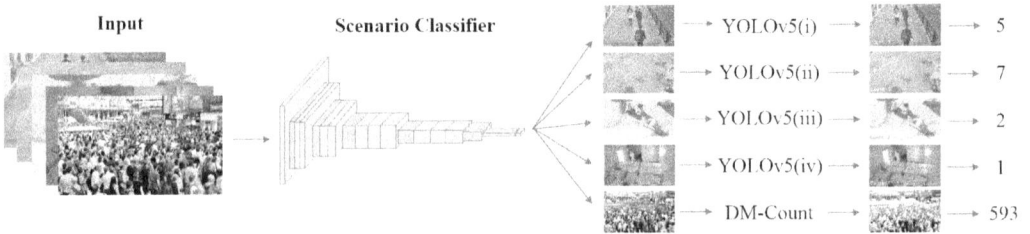

Figure 5.1  The workflow of our proposed paradigm. The input image is first passed to the scenario classifier to obtain a scenario label. Based on this label, the image is allocated to one of five person-counting models, which produce the final prediction on the number of persons in the image.

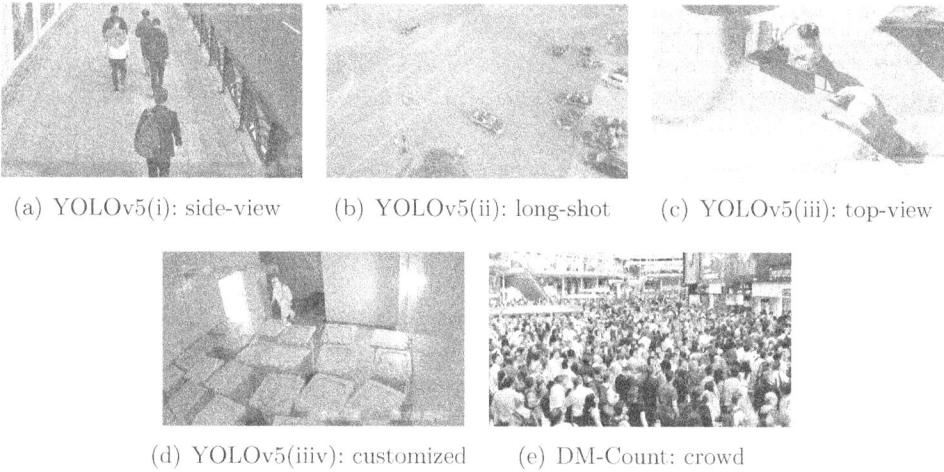

(a) YOLOv5(i): side-view     (b) YOLOv5(ii): long-shot     (c) YOLOv5(iii): top-view

(d) YOLOv5(iiiv): customized     (e) DM-Count: crowd

Figure 5.2   Examples of five scenarios: (i) side-view, where people are captured from a side view in mid-range, (ii) long-shot, where person bodies appear tiny, (iii) top-view, where people are captured from an overhead camera, (iv) customized, where people are wearing specific suits and (v) crowd, where numerous people appear in the image.

models, named YOLOv5(i) to YOLOv5(iv), are used for person detection and counting. For scenario (v), the DM-Count model generates a density map, with the final person count obtained by summing all pixel values in the map.

## 5.3.2   Scenario Classifier

The scenario classifier is a fundamental ResNet-50 [78] network. The input layer accepts an image with a size of $224 \times 224 \times 3$, followed by a $7 \times 7 \times 64$ convolution layer, a $3 \times 3$ max pooling layer, and four groups of bottleneck residual blocks. The first block contains three repeated units composed of three convolution layers with a kernel size of $1 \times 1 \times 64$, $3 \times 3 \times 64$, and $1 \times 1 \times 256$. The second block contains four repeated units composed of three convolution layers with a kernel size of $1 \times 1 \times 128$, $3 \times 3 \times 128$, and $1 \times 1 \times 512$. The third block contains six repeated units composed of three convolution layers with kernel sizes of $1 \times 1 \times 256$, $3 \times 3 \times 256$, and $1 \times 1 \times 1024$. The fourth block contains three repeated units composed of three convolution layers with kernel sizes of $1 \times 1 \times 512$, $3 \times 3 \times 512$, and $1 \times 1 \times 2048$. After all the convolution layers, an average pooling layer, a 5d fully connection layer, and a softmax layer are applied to predict the scenario category.

## 5.3.3   Person-Counting Module

The person-counting module consists of five models that are automatically selected based on the scenario label of the input image. These models include fine-tuned YOLOv5 and DM-Count architectures, each optimized for a specific scenario.

Four YOLOv5 models, referred to as YOLOv5(i) to YOLOv5(iv), are used in scenarios (i)-(iv) to detect and count the number of persons based on their bodies or heads. The YOLOv5 model consists of a backbone to aggregate and form image features at different granularity levels, a neck to mix and combine image features, and a head to consume features from the neck and predict the objectness score, class probability, and bounding box coordinates for each anchor box at multiple scales and aspect ratios. The anchor boxes are pre-defined shapes and sizes that cover different parts of the image.

For scenario (v), the DM-Count is used to estimate the density map for person counting. It adopts VGG-19 [204] as the backbone network and employs OT and TV loss instead of traditional Gaussian smoothing operations to avoid hurting the realness of the ground truth. Specifically, the OT loss measures the similarity between the predicted and ground truth density maps, and the TV loss is added to further enhance the smoothness of the predicted density map.

## 5.4 EXPERIMENTS

### 5.4.1 Dataset Overview

To train and validate our five person-counting models and scenario classifier, we constructed five augmentation datasets, which were also integrated to form the scenario classification dataset. Each augmentation dataset was collected from a specific scenario as defined in Section 5.1. The side-view dataset was mainly collected from public streets, parks, and offices et al., captured by cameras at a height of 3-5 meters, with each person annotated with a body bounding box. The long-shot dataset was collected from cameras placed at far distances from the subjects, such as surveillance cameras on highways, ferries, and squares et al., with each person also annotated with a body bounding box. The top-view dataset was collected from various sources, including a public overhead person detection dataset [171], with the annotations being head bounding boxes. The protective suit dataset, as an example of a customized scenario, was mostly collected from hospitals, health centers, and medical waste rooms, with the dataset characterized by data augmentation on people wearing protective suits and annotated with body bounding boxes. Lastly, the crowd-counting dataset was randomly selected from ShanghaiTech [260], UCF-CC50 [88], UCF-QNRF [89] and NWPU [230]. with each person marked with a dot on the head. For the scenario classification dataset, we combined all the images from the five augmentation datasets and labeled each image with a tag indicating the corresponding scenario. Additionally, we inferred the number of persons in each image from the raw annotations and calculated the maximum, minimum, and average values for each augmentation dataset and the integrated dataset. The statistics are presented in Table 5.1.

To train the person-counting models, each augmentation dataset is randomly divided into a training set and a validation set at a ratio of 8:2. As for the training of the scenario classifier, the division of the scenario classification dataset keeps consistent with that of the augmentation datasets for the convenience of joint network evaluations, which will be explained in Section 5.4.3.

TABLE 5.1   The statistics of five augmentation datasets and the integrated dataset. The columns from left to right show the dataset name, data scale, and the maximum, minimum, and average number of persons, respectively.

| Dataset | Scale | Max. | Min. | Avg. |
|---|---|---|---|---|
| Side-View | 5648 | 70 | 0 | 6.6 |
| Long-Shot | 5305 | 16 | 0 | 2.7 |
| Top-View | 4632 | 18 | 0 | 3.3 |
| Protective-Suit | 5904 | 6 | 0 | 1.8 |
| Crowd | 4834 | 12865 | 34 | 743.8 |
| Integrated | 26323 | 12865 | 0 | 138.1 |

## 5.4.2   Experiments Setup

The experiment projects are implemented on a workstation equipped with double Nvidia 3080Ti GPUs. The scenario classifier and five person-counting models are trained separately. For the scenario classifier, we utilize cross-entropy as the loss function and Adam [109] as the optimizer, with a learning rate of 0.001. The person-counting module is built on YOLOv5 and DM-Count. During the training phase of YOLOv5, the loss function is a combination of bounding box regression loss, object-ness loss, and classification loss, and is optimized using stochastic gradient descent (SGD) with an initial learning rate of 0.01 and momentum of 0.937. To train DM-Count, we use Adam with a learning rate of $1 \times 10^{-5}$ to optimize the overall loss function, which combines counting loss, OT loss, and TV loss.

During the inference phase of YOLOv5, all the generated proposals are first filtered by non-maximum suppression (NMS) with an intersection over union (IoU) threshold of 0.45 and confidence threshold of 0.6. The model then outputs the number of reserved bounding boxes as the final prediction. For DM-Count, the model produces a density map, and the sum of all the pixel values in the map is rounded to estimate the number of persons in the image.

## 5.4.3   Evaluations

The evaluations are designed to consider two aspects: the performance of the trained scenario classifier, and the effect of the scenario classifier on the overall network.

To evaluate the performance of our trained scenario classifier, we input all 5265 validation samples into the model and analyze the predictions. We adopt the one-vs-rest strategy to convert our multi-class problem into a series of binary ones. For each class, all the remaining classes are treated as the negative class, enabling us to use binary classification metrics such as precision, recall, and F1-score in a multi-class problem. Given the relatively balanced class distribution in our dataset, we use the macro average (the arithmetic mean of all metrics across classes) to reflect the performance of the scenario classifier on the entire validation dataset. The evaluation result and the confusion matrix are presented in Table 5.2 and Figure 5.3. It is observed that the scenario classifier produces generally fair predictions. However,

TABLE 5.2 The evaluation of the scenario classifier. The one-vs-rest strategy is used to calculate precision, recall, F1-score and support for each class, as well as the macro-average across all classes.

| Class | Prec. | Rec. | F1-score | Support |
|---|---|---|---|---|
| Side-View | 0.70 | 0.75 | 0.72 | 1130 |
| Long-Shot | 0.87 | 0.82 | 0.85 | 1061 |
| Top-View | 0.93 | 0.98 | 0.95 | 926 |
| Protective-Suit | 0.73 | 0.70 | 0.71 | 1181 |
| Crowd | 0.87 | 0.86 | 0.87 | 967 |
| Macro-Average | 0.81 | 0.81 | 0.81 | 5265 |

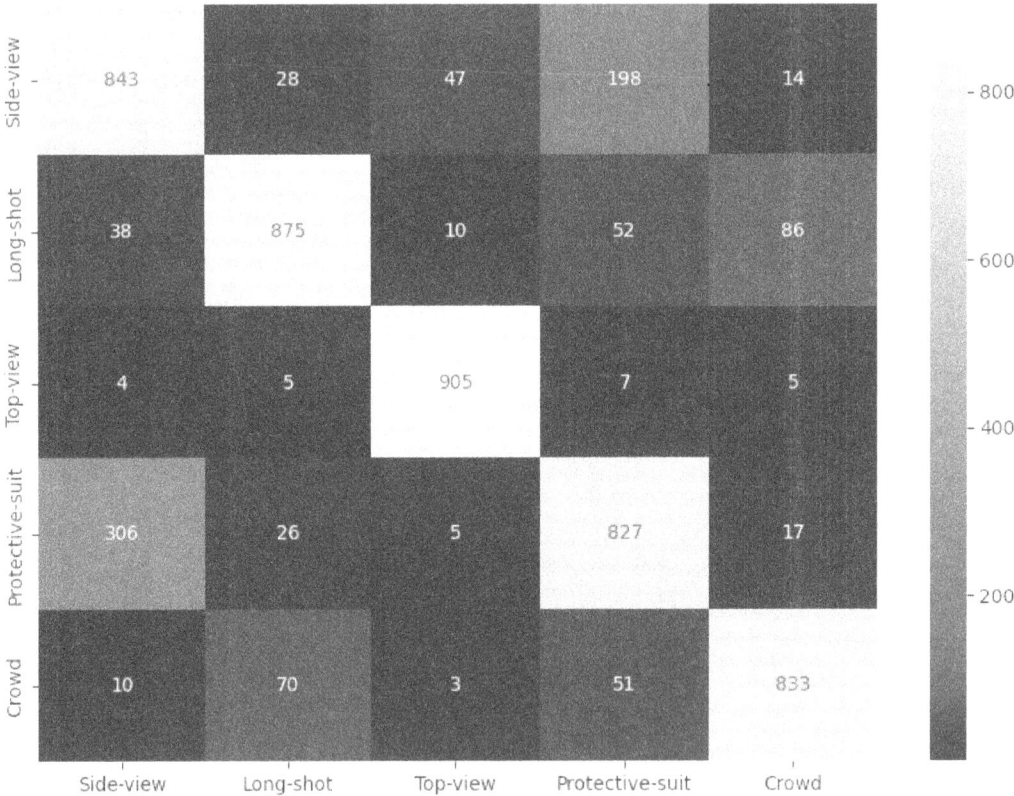

Figure 5.3 The confusion matrix of scenario classifier evaluation. Each grid in the matrix corresponds to a combination of ground truth and prediction, with the horizontal and vertical axes representing the ground truth and prediction, respectively.

some scenarios, such as side-view and protective suit, as well as long-shot and crowd, are more frequently mislabeled.

Regarding the effect of the scenario classifier, we compare the networks with and without the scenario classifier. When using the scenario classifier, the person-counting model is automatically selected based on the scenario classification result. Conversely, when not using the scenario classifier, the person-counting model is fixed to one of the five person-counting models. The evaluations are performed on the five augmentation datasets and the integrated dataset. We adopt MAE and RMSE to measure the difference between ground truth and the model's prediction. Recall that MAE and RMSE are calculated as follows:

$$\text{MAE} = \frac{1}{n} \sum_{i=1}^{n} |y_i - \hat{y}_i|$$

$$\text{RMSE} = \sqrt{\frac{1}{n} \sum_{i=1}^{n} (y_i - \hat{y}_i)^2}$$

(5.1)

where $n$ is the number of validation samples, $y_i$ and $\hat{y}_i$ are the ground truth and prediction of the $i$-th sample, respectively. As shown in Table 5.3, each person-counting model performs best on a specific scenario but produces higher deviation on other datasets. In comparison, with automatic model selection by the scenario classifier, our network achieves a great balance on the integrated dataset.

### 5.4.4 Visualization Results

As illustrated in Figure 5.4, we visualize the inference results of the person-counting models on several samples to demonstrate the performance of our proposed paradigm. Specifically, we select two validation images from each augmentation dataset and apply all the person-counting models for prediction. For YOLOv5(i), (ii), and (iv), we render body bounding boxes on the image, while for YOLOv5(iii), we render head bounding boxes. For the DM-Count model, we first blur the estimated density map with a Gaussian kernel with the size of $5 \times 5$ to generate a heatmap, which is then overlaid onto the original image for better visualization. We also remark the counting number on the left-bottom of each output image. From these results, we observe that each person-counting model shows its strengths and weaknesses in specific scenarios. Moreover, the existence of a scenario classifier raises the possibility for the input image to be processed by the appropriate person-counting model.

## 5.5 CONCLUSION

In this work, we propose an AI paradigm specially designed for person-counting tasks, which takes into account the scenario in which the image is captured. The proposed architecture consists of a scenario classifier and a person-counting module containing four YOLOv5 models and a DM-Count model, each fine-tuned for a specific scenario. The scenario classifier has proved effective in allocating the input image to one of five person-counting models based on its scenario label. The YOLOv5-based models

TABLE 5.3 Comparisons on the networks with and without the scenario classifier. The first five rows report the results of fixed person-counting models without using the scenario classifier, while the last row shows the performance of the model automatically selected by the scenario classifier. The experiments are conducted on five separate datasets and the integrated dataset, using MAE and RMSE as evaluation metrics.

| Model | Side-View | | Long-Shot | | Top-View | | Protective-Suit | | Crowd | | Integrated | |
|---|---|---|---|---|---|---|---|---|---|---|---|---|
| | MAE | RMSE | MAE | RMSE | MAE | RMSE | MAE | RMSE | MAE | RMSE | MAE | RMSE |
| YOLOv5(i) | **0.39** | **0.94** | 1.02 | 1.44 | 1.24 | 1.57 | 0.76 | 1.02 | 613.6 | 721.3 | 113.4 | 309.1 |
| YOLOv5(ii) | 0.68 | 1.13 | **0.36** | **0.80** | 1.21 | 1.48 | 0.91 | 1.12 | 583.0 | 622.5 | 107.7 | 266.8 |
| YOLOv5(iii) | 1.88 | 2.31 | 1.38 | 1.61 | **0.44** | **0.89** | 1.42 | 1.69 | 624.7 | 736.4 | 115.8 | 315.6 |
| YOLOv5(iv) | 0.44 | 0.91 | 1.09 | 1.49 | 1.26 | 1.57 | **0.28** | **0.69** | 616.2 | 725.8 | 113.8 | 311.1 |
| DM-Count | 7.56 | 10.37 | 13.23 | 18.63 | 6.45 | 8.77 | 6.84 | 10.1 | **96.4** | **153.8** | 24.7 | **66.9** |
| **Automatic** | 0.41 | 0.93 | 0.58 | 1.01 | 0.66 | 1.23 | 0.54 | 0.92 | 108.6 | 168.1 | **20.4** | 72.0 |

| Image | YOLOv5(i) | YOLOv5(ii) | YOLOv5(iii) | YOLOv5(iv) | DM-Count | Sample info. |
|-------|-----------|------------|-------------|------------|----------|--------------|
| (a) | | | | | | scenario label = 0<br>scenario pred. = 0<br>ground truth = 5<br>prediction = 5 |
| (b) | | | | | | scenario label = 0<br>scenario pred. = 0<br>ground truth = 1<br>prediction = 1 |
| (c) | | | | | | scenario label = 1<br>scenario pred. = 1<br>ground truth = 1<br>prediction = 1 |
| (d) | | | | | | scenario label = 1<br>scenario pred. = 1<br>ground truth = 8<br>prediction = 7 |
| (e) | | | | | | scenario label = 2<br>scenario pred. = 2<br>ground truth = 10<br>prediction = 9 |
| (f) | | | | | | scenario label = 2<br>scenario pred. = 2<br>ground truth = 2<br>prediction = 2 |
| (g) | | | | | | scenario label = 3<br>scenario pred. = 3<br>ground truth = 1<br>prediction = 1 |
| (h) | | | | | | scenario label = 3<br>scenario pred. = 3<br>ground truth = 3<br>prediction = 3 |
| (i) | | | | | | scenario label = 4<br>scenario pred. = 4<br>ground truth = 584<br>prediction = 593 |
| (j) | | | | | | scenario label = 4<br>scenario pred. = 1<br>ground truth = 196<br>prediction = 99 |

Figure 5.4 Visualization results of our experiments. The input images are picked from five augmentation datasets and each image is processed by all the five person-counting models. For YOLOv5 models, we render the detected body/head bounding boxes on the image. For DM-Count model, we overlay the generated heatmap on the original image. And the counting result is remarked on each output image. The last column indicates the actual and the predicted scenario labels of the image, along with the ground truth and final prediction of person counting for each image. In case (a)-(i), the scenario classifier successfully allocates the input image to the correct person-counting model, while in case (j), the input image is mislabeled thus passed to a model that is not the most appropriate.

count the persons by detecting bodies or heads and counting the number of bounding boxes, while the DM-Count produces an estimation by generating a density map and summing up all the pixel values. Our paradigm outperforms any single predetermined model on the integrated validation dataset, demonstrating its generalization in various scenarios.

# CHAPTER 6

# Operation Procedure Detection Paradigm for Noncompliant Operations Detection of Oil Unloading

## 6.1 INTRODUCTION

In industrial production, the importance of operators' compliance with operational procedures cannot be ignored. It not only guarantees the integrity and reliability of products but also plays a vital role in enhancing production efficiency and reducing safety risks. Consider automotive assembly, where even the slightest deviation from standard operating procedures can lead to defects in the final product. Similarly, in oil unloading operations, any non-compliance could potentially trigger safety accidents, posing a grave threat to personnel and property. Therefore, it is crucial to conduct real-time compliance monitoring of operational procedures. However, traditional compliance detection methods heavily rely on manual or remote video inspections, which are inefficient and unable to promptly address potential risks.

To address this challenge, we have developed an AI-based paradigm for operation procedure compliance detection. As shown in Figure 6.1, our paradigm utilizes visual detection and tracking algorithms to analyze operational surveillance videos in real-time, accurately tracing the movement trajectories of operators and equipment on-site. When an interaction is sustained over a certain duration between operators and equipment trajectories, we can determine that the operator is operating the equipment by trajectory fusion decision-making. Subsequently, we assess the compliance of the operation based on established procedural rules. Our paradigm is versatile and can be applied in any industrial operation compliance detection scenario.

As a highly risky operation, oil unloading requires strict adherence to standardized procedures. Non-compliance significantly increases the danger of fires or explosions. Therefore, we will take the oil unloading operation as an example to provide a detailed explanation of our proposed paradigm.

DOI: 10.1201/9781003644972-6

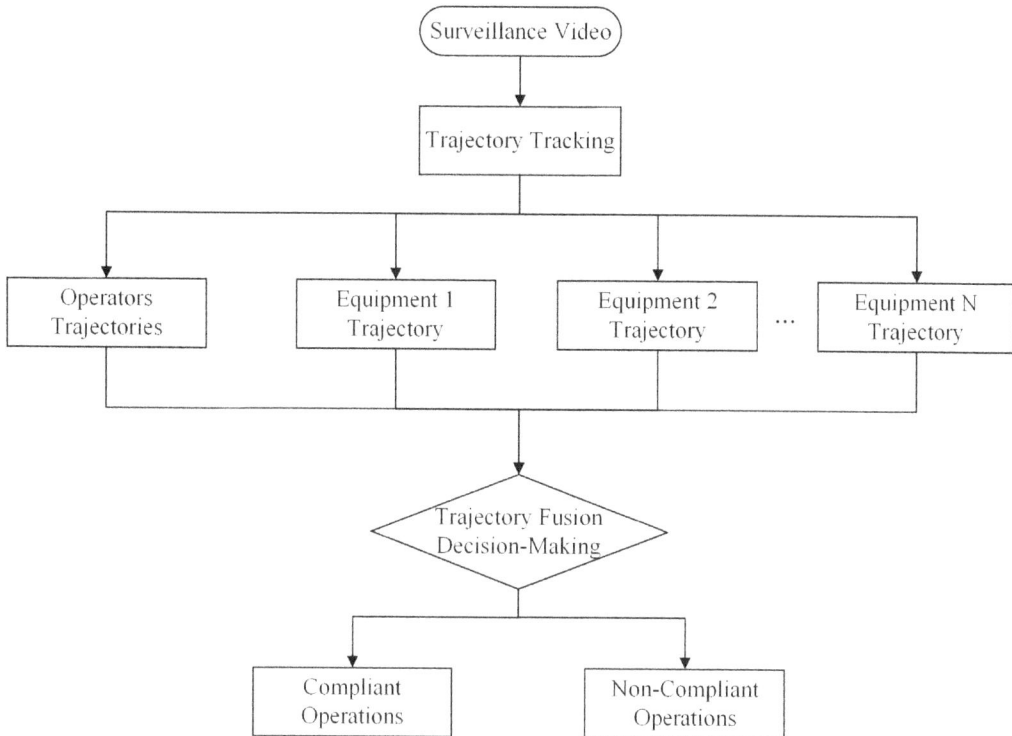

Figure 6.1  The framework of our operation procedure compliance detection paradigm.

## 6.2  RELATED WORK

In the field of operation procedure compliance detection, early studies primarily relied on traditional sensor-based technologies. For instance, smart gloves equipped with embedded sensors were used to capture hand movements, providing foundational data to evaluate whether operators followed prescribed procedures [120]. However, these methods were constrained by physical installation requirements and signal interference, making it difficult to obtain a comprehensive view of the operation process and limiting their applicability in complex scenarios.

With the rapid development of computer vision and deep learning technologies, video analysis has emerged as one of the primary approaches for compliance detection. Convolutional Neural Networks (CNNs) [130, 238] are extensively utilized in object recognition and behavior pattern analysis. For example, the YOLO object detection algorithm enables real-time monitoring of operator-equipment interactions [179, 96, 87]. By accurately detecting and tracking multiple objects in videos, even in complex backgrounds, YOLO provides detailed observation of operational processes while maintaining high frame rates, significantly improving efficiency and accuracy.

More recently, pose estimation-based methods have also gained widespread attention. OpenPose [20, 202, 21, 232], a real-time multi-person 2D pose estimation tool, captures key human body points such as joints and limb positions to construct a complete body model. This approach is not only effective for recognizing static poses but also excels in analyzing complex dynamic actions, providing more intuitive and detailed visual feedback for compliance detection.

Building upon these studies, we propose an AI-based paradigm that leverages visual detection and trajectory fusion decision-making mechanisms to achieve real-time monitoring and compliance evaluation of operational processes.

## 6.3 METHOD

In this section, we will elaborate on the details of our paradigm through oil unloading operation compliance detection. We initially introduce the key steps and corresponding operational standards in oil unloading. Then, we detail how our paradigm detects and judges whether the operations of these key steps are compliant.

### 6.3.1 Key Steps of Oil Unloading Operations

Oil unloading operations consist of three key steps: connecting static wires, deploying safety equipment, and connecting oil unloading pipes. As shown in Figure 6.2 (a), operators are required to attach the static wires from the static grounding pile to the tanker in the connecting static wires step. Figure 6.2 (b) depicts the arrangement of safety equipment. In this step, operators are required to place safety equipment in specific locations. Figure 6.2 (c) specifically illustrates the process of connecting the oil unloading pipes. In this step, operators need to attach both ends of the pipe to the unloading port and the underground tank port, respectively. Notably, the

(a) connecting static wires

(b) deploying safety equipment

(c) connecting oil unloading pipes

Figure 6.2　The key steps of oil unloading operations. The boxes represent the operators and corresponding equipment, respectively.

completion of the oil unloading pipe connection signifies the imminent start of the unloading process. Before connecting the unloading pipes, operators must first ensure the electrostatic grounding cable is connected to dissipate static electricity, and set up necessary safety equipment such as warning signs and fire extinguishers.

### 6.3.2  Operation Procedure Compliance Detection

As shown in Figure 6.1, our paradigm for operation procedure compliance detection is comprised of two main steps: trajectory tracking and trajectory fusion decision-making. The trajectory tracking step involves detecting and tracking the movement trajectories of operators and equipment. In the trajectory fusion decision-making step, we analyze these trajectories to determine whether the operators are indeed operating the equipment and further assess the compliance of their operations against the established procedural rules.

#### 6.3.2.1  Trajectory Tracking

We employ YOLOv8 [99] to detect operators and oil unloading equipment, and utilize the ByteTrack [259] tracking algorithm to monitor the movement trajectories of both operators and the equipment in real-time.

As shown in Figure 6.3, the YOLOv8 model is comprised of the Backbone, Neck, and Head. The Backbone utilizes an enhanced CSPDarknet53 [13] architecture. The Neck incorporates PAFPN [138] modules to achieve multi-scale feature fusion. The Head adopts a decoupled structure, separating classification and detection tasks, and transitions from an Anchor-Based [258] to an Anchor-Free [264] approach. Overall, YOLOv8 is designed for efficiency, ensuring high-precision detection.

The ByteTrack algorithm is an efficient and accurate multi-object tracking method that operates without requiring additional training and relies solely on object detection results. ByteTrack initially classifies the detection boxes generated by detection algorithms such as YOLOv8 into high-score and low-score groups, giving priority to matching high-score detection boxes with existing tracking trajectories. For trajectories that fail to match successfully, ByteTrack attempts to re-match them using low-score detection boxes, effectively handling trajectory interruptions caused

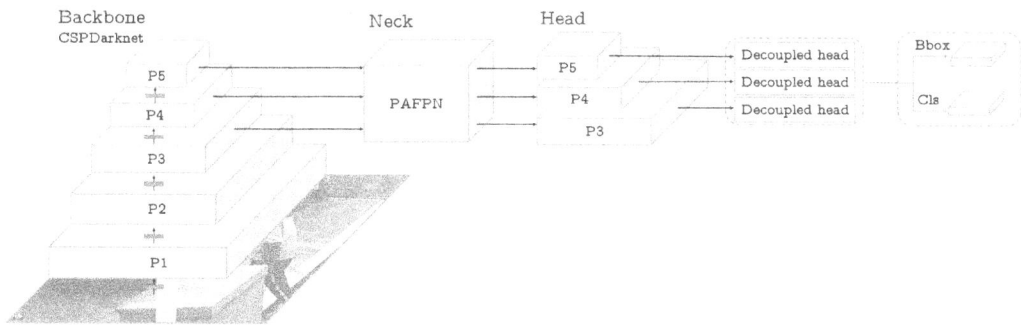

Figure 6.3  YOLOv8 network architecture overview.

by occlusions or other factors. Additionally, ByteTrack further enhances the accuracy and robustness of tracking by predicting the position of objects in subsequent frames and optimizing the matching process of detection boxes.

### 6.3.2.2  *Trajectory Fusion Decision-Making*

During trajectory tracking, we continuously collect and analyze trajectory data to determine whether the motion paths of operators and equipment are fused in spatio-temporal. A sustained fusion intersection indicates that the operator is operating the equipment.

Figure 6.4 provides an example of trajectory fusion decision-making. Specifically, when we detect that the movement path of an operator merges with that of the grounding wire and this fusion lasts for a period, we can infer that the operator is engaging in connecting the grounding wire. Once the presence of a grounding clamp is detected on the tanker, it can be confirmed that the connection of the grounding wire has been completed. Similarly, the arrangement of safety equipment and the connection of the unloading pipe are also confirmed through the same trajectory analysis.

Based on the above trajectory fusion analysis, we can further evaluate whether the operators' operational behaviors adhere to the prescribed standards. For instance, after confirming the completion of the unloading pipe connection, we will check whether the static wire connection and the safety equipment arrangement have been executed. In the event of any incomplete operation, an immediate non-compliance alert will be triggered to ensure the safe execution of oil unloading operations at the gas station.

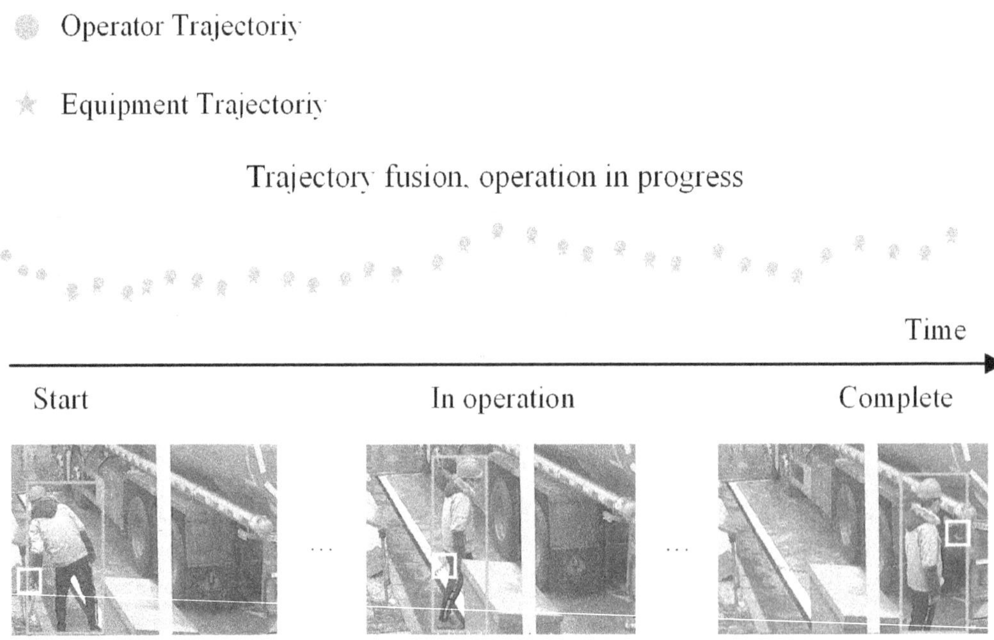

Figure 6.4   Trajectory fusion decision-making: connecting static wires operation.

## 6.4 EXPERIMENTS

In this section, we validate the effectiveness of our operational procedure compliance detection paradigm through an experiment on oil unloading compliance detection. Specifically, we conduct a detailed analysis of the accuracy and effectiveness of the two key steps in our paradigm: trajectory tracking and trajectory fusion decision-making.

### 6.4.1 Dataset

We have collected 1000 surveillance videos of oil unloading operations from 300 gas stations under various weather and lighting conditions. These videos were randomly divided into a training set and a test set at a ratio of 9:1. We extracted frames from the 900 videos in the training set at a 5-second interval, resulting in a total of 54,328 images. We annotated key targets such as humans, tankers, and unloading equipment in these images for subsequent model training.

For the 100 videos in the test set, we also extracted frames and performed annotations at a 5-second interval to validate the performance of trajectory tracking. Additionally, we invited professional operators to manually assess the 100 oil unloading test videos, recording whether the operations were compliant, to verify the accuracy of the trajectory fusion decision-making.

### 6.4.2 Training

We conducted model training for object detection using the default parameter configuration provided by the official YOLOv8 implementation.

### 6.4.3 Experiment of Trajectory Tracking

We adopted the MOTA (Multiple Object Tracking Accuracy) [152] and IDF1 (ID F1 Score) [184] metrics to evaluate the accuracy and performance of our trajectory tracking approach. MOTA, as a comprehensive indicator, takes into account false positives (FP), false negatives (FN), and identity switches (IDS), providing a measure of the overall accuracy in multi-object tracking tasks. It is formulated as follows:

$$\text{MOTA} = 1 - \frac{\sum_t \left( \text{FP}_t + \text{FN}_t + \text{IDS}_t \right)}{\sum_t \text{GT}_t} \tag{6.1}$$

where $\text{FP}_t$, $\text{FN}_t$, $\text{IDS}_t$ represent the number of false positives, false negatives, and identity switches, respectively, at time t, and $\text{GT}_t$ denotes the number of ground truth objects at time t.

IDF1 specifically focuses on assessing the performance of identity recognition, incorporating true positives (TP), FP, and FN into its evaluation. The IDF1 score is defined as:

$$\text{IDF1} = \frac{2 \cdot \text{IDTP}}{2 \cdot \text{IDTP} + \text{IDFP} + \text{IDFN}} \tag{6.2}$$

TABLE 6.1   Performance evaluation of trajectory tracking.

| Metric | Result |
|--------|--------|
| MOTA | 0.75 |
| IDF1 | 0.82 |

where IDTP represents the number of true positive identifications, IDFP is the number of false positive identifications, and IDFN denotes the number of false negative identifications.

Table 6.1 presents the results of our tracking performance evaluation. As can be seen from the table, our trajectory tracking method achieved a score of 0.75 on the MOTA metric, indicating that it can effectively reduce the occurrence of false positives and false negatives in multi-object tracking tasks while minimizing identity switches, thus ensuring high overall tracking accuracy. Furthermore, with a score of 0.82 on the IDF1 metric, our method demonstrates excellent performance in identity recognition, accurately identifying and distinguishing different target objects. The results demonstrate that our trajectory tracking method exhibits high accuracy and remarkable identity recognition capabilities in multi-object tracking tasks, laying a solid foundation for subsequent trajectory fusion decision-making.

### 6.4.4   Experiment of Trajectory Fusion Decision-Making

Based on the trajectory tracking, we experimented to validate the effectiveness of our trajectory fusion decision-making. Table 6.2 compares the results of trajectory fusion decision-making using our paradigm with those of manual judgments. As can be seen, our results exhibit a remarkably high degree of consistency with the manual judgments in both compliant and non-compliant operation categories.

In the complaint category, where the oil unloading operations followed the prescribed procedures, our method achieved a consistency rate of 95.85%. This high consistency indicates that our automatic decision-making paradigm is capable of accurately identifying compliant operations, minimizing false positives. In the non-compliant category, where deviations from the standard procedures occurred, our method achieved an even higher consistency rate of 96.11%. This result demonstrates the robustness of our method in detecting non-compliant operations, ensuring that potential safety hazards are not overlooked.

TABLE 6.2   Consistency evaluation of compliant and non-compliant operations between our paradigm and professional operator.

| Category | Number of Operations | Consistency Rate |
|----------|---------------------|------------------|
| Compliant | 217 | 95.85% |
| Non-Compliant | 103 | 96.11% |

The high consistency rates achieved in both categories validate the effectiveness of our trajectory fusion decision-making method in determining the compliance of oil unloading operations. This, in turn, verifies the practicality and reliability of our proposed operational procedure compliance detection paradigm.

## 6.5 CONCLUSION

In this chapter, we propose an AI-based paradigm for operation procedure compliance detection that leverages visual detection and tracking algorithms to precisely trace the movement trajectories of operators and equipment. By integrating trajectory fusion and decision-making techniques, our paradigm can effectively judge whether operators adhere to the prescribed equipment operation procedures. Experiments conducted on the compliance detection of oil unloading operations have achieved a detection rate of 96.11% for non-compliant operations and validated the effectiveness of our paradigm. Furthermore, our paradigm is not limited to oil unloading operations but is equally suitable for compliance detection in a diverse range of operational processes, offering a versatile solution for enhancing safety and efficiency across various industrial production.

# CHAPTER 7

# Object Measurement Paradigm for Length Measurement of Iron Chains

## 7.1 INTRODUCTION

Object measurement is a critical component in various industrial and scientific applications, as accurate dimensional measurement significantly impacts quality control and performance evaluation. Currently, manual measurements predominantly rely on highly precise instruments such as calipers and micrometers. Despite strict adherence to operational protocols by personnel, these instruments are inherently prone to unpredictable inaccuracies due to the issue of human error within manual operations.

There have been several explorations into automated measurement methods, including mechanical-touching, laser scanning, and 3D reconstruction methods. While the mechanical-touching method offers high measurement accuracy, it also comes with significant drawbacks, particularly with regard to its hardware components that can be notably expensive and challenging to repair or replace when damaged. Moreover, this method is not well-suited for irregularly shaped objects. Similarly, laser scanning shares similar limitations with the mechanical-touching approach; it tends to become less efficient when dealing with larger objects. 3D reconstruction, as exemplified by [92], presents an innovative solution to this problem. It allows for the accurate digital recreation of objects, making it effective for measuring irregular items. However, the precision and quality of the reconstructions are sometimes insufficient to meet the stringent requirements of industrial measurement tasks.

In this chapter, we propose an AI-driven paradigm for industrial object measurement. We employ deep learning algorithms to detect the objects to be measured and pinpoint their respective $2D$ keypoints. Subsequently, depth information obtained from the RGBD camera is used to transform these $2D$ keypoints into their corresponding $3D$ coordinates. Next, we utilize custom-designed geometric correlation rules to accurately calculate the lengths of the objects.

DOI: 10.1201/9781003644972-7

## 7.2 RELATED WORK

### 7.2.1 Mechanical-Touching Method

The mechanical-touching method is a common approach that involves closely contacting objects with sensors. Typically, instruments such as calipers and micrometers all belong to the category of mechanical-touching method. However, these tools are human-operated, which can introduce subjectivity and reduce objectivity. In contrast, numerous auto-mechanical measurement techniques [196, 210] have been widely adopted in industry. These methods are notably more reliable, enabling efficient and rapid digitization of processes. Nevertheless, these auto-mechanical measurement techniques come with certain limitations, such as the requirement for complex and costly devices.

### 7.2.2 Laser Scanning

Laser scanning represents a sophisticated approach to object measurement, which typically comprises a laser emitter, a receiver, and a control circuit [234]. This technique is renowned for achieving high precision alongside an extensive measurement range. The accuracy of laser scanning generally reaches within millimeters, making it suitable for detailed and accurate measurements. Operationally, the process involves emitting either a single laser beam or a series of pulses towards the target object. The receiver then captures the reflected signals from the object. A crucial component of this system is its high-precision timing mechanism, which accurately measures the duration it takes for the pulse to travel to the target and back. Based on this time-of-flight information, precise distance calculations can be made. Despite these advantages, laser scanning comes with notable limitations [18]. The method requires a well-controlled environment; environmental factors such as dust, smoke, or raindrops can significantly degrade the precision and reliability of the measurements.

### 7.2.3 3D Reconstruction

The 3D reconstruction technique is primarily employed for performing high-precision three-dimensional modeling of objects [267]. This process facilitates the calculation of object dimensions by utilizing the reconstructed 3D information. The accuracy of these calculations hinges directly on the precision of the 3D reconstruction; higher reconstruction accuracy leads to more precise measurements.

Various 3D reconstruction methods can be broadly categorized into two types based on the input data format: those based on depth information [86] and those based on images [75]. Depth-based methods typically use a 3D sensor to capture depth maps, which, when combined with the intrinsic parameters of the camera, can be converted into 3D point cloud data for further analysis. Image-based methods, on the other hand, involve extracting geometric and depth information from multiple images. By analyzing one or more images, the intrinsic and extrinsic parameters of the camera are estimated using geometric constraints. Subsequently, Multi-View Stereo (MVS) or Structure-from-Motion (SfM) techniques [195] are applied to analyze the point clouds of the objects.

In the context of 3D reconstruction, manual rules can be utilized to identify key points essential for calculating geometric dimensions. These rules leverage discernible features such as textures and geometric characteristics to efficiently pinpoint relevant keypoints on the models.

3D reconstruction technology has become an essential field with significant potential. Measurement based on 3D reconstruction offers advantages such as speed, simplicity, and realism, enabling effective digital reconstruction of objects. Despite its impressive performance in various applications, challenges persist, including computationally intensive processes, performance lags, inconsistent reconstruction stability, and the requirement for expensive hardware.

### 7.2.4 Keypoint Detection Method Based on Manual Rules

Traditional image processing provides powerful methods for analyzing object information, including feature extraction [157], corner detection [124], and line feature extraction [57]. These techniques find broad applications across various fields. However, their effectiveness is highly dependent on the design of sophisticated extractors and requires human-defined rules, which become increasingly complex as precision demands rise. In scenarios where scene parameters change—such as variations in lighting or background conditions—unexpected errors can easily occur.

For instance, changes in ambient lighting can significantly affect the accuracy of feature extraction, while varying backgrounds may introduce noise that complicates corner or line detection. Therefore, although traditional image processing methods are potent tools for specific conditions, they face limitations when applied to dynamic or less controlled environments.

### 7.2.5 Keypoints Detection Method Based on Deep Learning

Deep learning networks have achieved significant breakthroughs in the field of computer vision. Among the state-of-the-art object detection methods, one-stage detectors such as YOLO [179, 180, 181, 98], and two-stage methodologies like Faster R-CNN [182]. Built upon these foundations, keypoints detection algorithms [198, 24] have also gained widespread adoption across various applications, including face detection [220], human pose estimation [6], gesture recognition [91], and more.

In terms of keypoint detection, there are primarily two approaches:

- Coordinate-based methods like DeepPose [215] directly regress keypoints from pixel coordinates. Despite offering a relatively rapid inference rate, the guaranteed accuracy of predictions is limited.

- Spatial distribution-based (heatmap-based) methods [240] have gradually gained widespread acceptance. This technique represents keypoints' coordinates as heatmaps and assigns a confidence score to each position. By leveraging the predicted heatmap, keypoint locations can be further extracted. Due to their ability to better preserve image information, these methods are more suitable for CNN design compared to other alternatives, and they generally exhibit higher accuracy than direct coordinate regression.

## 7.3   THE PROPOSED AI PARADIGM

In this section, we provide a detailed introduction to the application-driven AI paradigm for industrial object measurement. This paradigm uses deep learning algorithms to detect keypoints on the target object. Subsequently, geometric constraints are applied through predefined manual rules to calculate inter-keypoint distances. These calculations ultimately determine the total length of the target object.

The workflow of our paradigm is illustrated in Figure 7.1 and is divided into two main stages. The first stage employs deep learning algorithms for 2D keypoint estimation, which includes both object detection and keypoint localization. In the second stage, the object size is calculated by applying manual rules that integrate depth information as a key factor. This depth data enables the transformation of keypoints from 2D to 3D space. By combining this spatial information with the known geometric properties of the target object, the size parameters, such as the chain length, can be accurately computed.

### 7.3.1   2D Keypoints Estimation Based on Deep Learning Methods

We first detect the target object, specifically the iron chain, using a pair of RGBD cameras positioned above and in front of the chain to capture images from two angles, categorizing the chain's hoops into two types based on these viewpoints as illustrated in Figure 7.2. Following detection, we perform keypoint estimation on the identified areas, selecting the endpoints at both extremities of the chain as keypoints to accurately represent its size, with the estimated results shown as blue-marked points in the right-side images of Figure 7.2(a) and Figure 7.2(b).

#### 7.3.1.1   *Iron Chain Detection*

We chose YOLOv5 [98] for executing the detection tasks due to its excellent detection performance and high speed, making it well-suited for industrial applications. The bounding boxes generated by YOLOv5 require preprocessing before proceeding to the next stage. First, we standardize the aspect ratios of these boxes by uniformly resizing

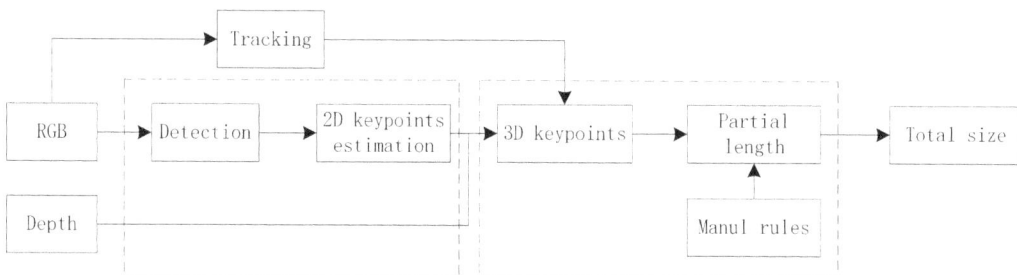

Figure 7.1   The workflow of our AI Paradigm, includes two main stages. Stage one leverages deep learning for object detection and 2D keypoint estimation. In stage two, these 2D keypoints are transformed into a 3D representation, and the length of the objects is calculated based on established manual rules.

(a) front view

(b) top view

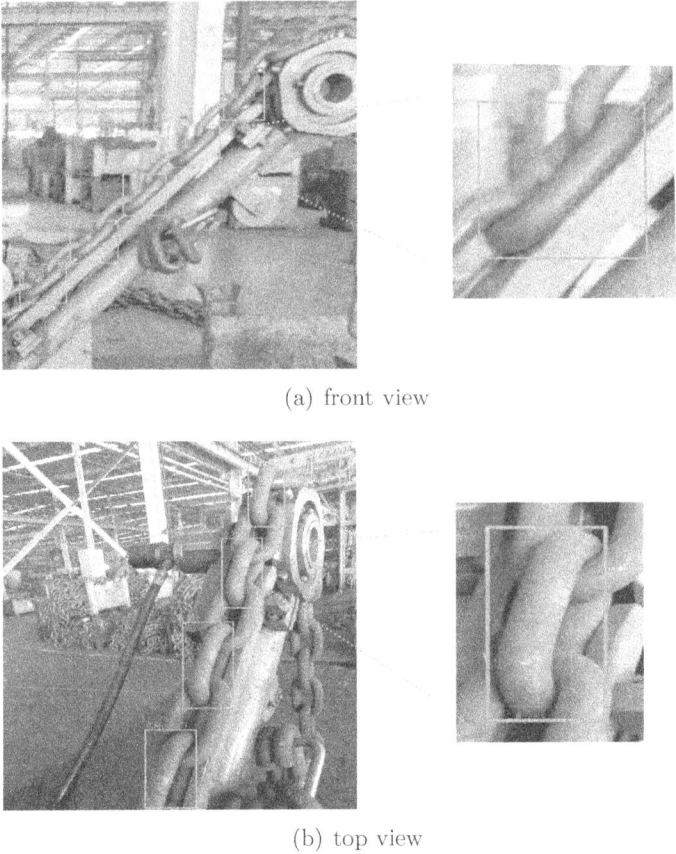

Figure 7.2 Iron chain detection and keypoints detection. The left images in Figure 7.2(a) and 7.2(b) demonstrate the differences between these hoop types. The estimated results from this task are depicted by the blue points shown in the right images of Figure 7.2(a) and 7.2(b).

either their width or height. Then, we slightly enlarge the boxes to include some of the surrounding background, as keypoints may sometimes lie near the boundaries, which could otherwise hinder their accurate detection in subsequent steps.

### 7.3.1.2 Keypoints Estimation

For the extraction of keypoints within the detected bounding boxes, we adopt the heatmap and offset network structure from [164]. We feed cropped images into a ResNet-101 network to generate potential heatmaps and offsets. These features consist of three dimensions: one dimension indicates whether a pixel represents a keypoint, while the other two dimensions provide coordinate biases in the form of x and y values. The final output comprises a total of $3K$ channels, with $K$ denoting the number of keypoints.

In [164], Papandreou et al. used heatmaps with binary values (0 or 1), which simplified and expedited the training process. However, for object measurement tasks where accuracy is paramount, especially in practical production applications, this approach is insufficient. Therefore, we configure our heatmap values to fall within the range $(0, 1.0)$, where each value signifies the confidence level. Specifically, we assign pixels within a radius $R$ around keypoints varying values based on their distances $r$ from the keypoints. This computation is performed as follows:

$$r = \sqrt{(x - x_0)^2 + (y - y_0)^2}, \tag{7.1}$$

$$f(x, y) = \begin{cases} 1 - \frac{r}{R}, r < R \\ 0, r \geq R \end{cases}, \tag{7.2}$$

in which $x_0$ and $y_0$ denote the coordinates of a keypoint, while $x$ and $y$ represent the corresponding pixels in the heatmap. The heatmap will contain $K$ distinct groups, and the classification problem will be addressed for each keypoint independently. Moreover, apart from the heatmap, we predict an offset for every pixel within the range, which is represented by a vector pointing towards the keypoints. Consequently, there will be $K$ separate vectors as well.

We use ResNet [78] as the backbone with two convolutional heads. The losses are defined as:

$$L_h(x) = \begin{cases} \sum_{k=0}^{K} \sum_{i=0}^{W} \sum_{j=0}^{H} (C_{x_i, y_i} * f(x_i, y_i) - f(x_k, y_k)), r < R \\ 0, else \end{cases} \tag{7.3}$$

$$L_o(x) = \begin{cases} \sum_{k=0}^{K} \sum_{r<R}^{R} H(||F_k(x_i, y_j) - T_k(x_i, y_j))||), r < R \\ 0, else \end{cases} \tag{7.4}$$

$$L(x) = \lambda_h * L_h(x) + \lambda_o * L_o(x). \tag{7.5}$$

In the above equations, $L_h(x)$ and $L_o(x)$ denote the losses for heatmaps and off-sets. $f(x_i, y_j)$ represents the coordinates within the heatmaps, while $C_{x_i, y_j}$ is the corresponding confidence. The Huber robust loss function is denoted by $H(u)$. $F_k(x_i, y_j)$ signifies the predicted vector pointing from pixel $i$ to keypoint $k$, with $T_k(x_i, y_j)$ being the ground truth vector. The final loss function is defined in Eq. 7.5. $\lambda_h$ and $\lambda_o$ are scalar coefficients that serve to balance the two loss terms and expedite convergence.

## 7.3.2 Size Calculation Based on Manual Methods

### 7.3.2.1 3D Keypoints Transforming

Through the aforementioned network, we obtain the corresponding 2D keypoint co-ordinates. However, these keypoints are initially located in the coordinate system of the cropped image and must be transformed into the original image's coordinate system using the bounding box coordinates. Subsequently, the 3D coordinates of the keypoints are derived from their 2D counterparts by leveraging depth information.

The transformation process is defined by the following equations:

$$\begin{cases} x = \frac{(u-c_x)}{f_x} \cdot z \\ y = \frac{(v-c_y)}{f_y} \cdot z \\ z = d(u,v) \end{cases} \tag{7.6}$$

Here, $(u,v)$ are the 2D coordinates of the keypoints, $(c_x, c_y)$ are the principal point coordinates, $(f_x, f_y)$ are the focal lengths, and $d(u,v)$ represents the depth value at the keypoint location.

Before converting the 2D keypoints to 3D, camera calibration is essential. Calibrated parameters correct lens distortion and establish precise object locations within the scene, enabling measurements in world units. Higher accuracy in calibrated parameters leads to more precise 3D point coordinates. Using Eq. 7.6, we transform the 2D keypoint coordinates into 3D coordinates, allowing for accurate calculation of the iron chain's dimensions based on predefined rules.

### 7.3.2.2 Size Calculation

Based on the regular geometry of the chain, we identify pairs of corresponding keypoints to represent the length of individual hoops. These lengths serve as pivotal parameters for calculating the entire chain's length. Given potential occlusions, a single camera perspective may not suffice for directly measuring hoop lengths. Therefore, we use two calibrated cameras to capture both frontal and side views of the chain, as shown in Figure 7.3.

Initially, these cameras are calibrated using the method proposed by Cui et al. [39]. The calibration process yields a transformation matrix that converts coordinates from camera space to world coordinates. This matrix is then applied to transform the 2D

(a)                                                                    (b)

Figure 7.3   Different views of hoops. The overlap between adjacent hoops precludes a straightforward summation of each hoop's length to determine the total chain length.

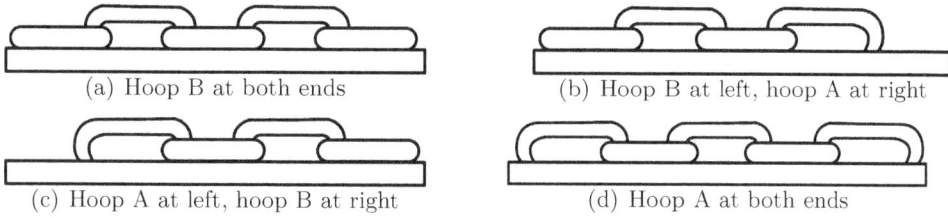

(a) Hoop B at both ends    (b) Hoop B at left, hoop A at right

(c) Hoop A at left, hoop B at right    (d) Hoop A at both ends

**Figure 7.4**  Different placement of hoops.

keypoint coordinates obtained from each view into 3D world coordinates, ensuring accurate measurements despite occlusions.

As depicted in Figure 7.3, the overlap between adjacent hoops precludes a straightforward summation of each hoop's length to determine the total chain length. Hence, we employ several manual rules for further calculations.

The situation can be approximately categorized into four distinct cases based on the preliminary insights presented in Figure 7.4, with the corresponding calculations carried out as follows:

$$
L = \begin{cases}
M_{r_0}^{t_0} - M_{l_0}^{t_0} + \sum\limits_{i=1}^{n}(M_{r_i}^{t_i} - M_{r_{i-1}}^{t_i}) \\[2mm]
\sum\limits_{i=0}^{n}(M_{l_i}^{t_i} - M_{l_{i-1}}^{t_i}) + N_{r_n}^{t_n} - M_{l_n}^{t_n} \\[2mm]
M_{r_0}^{t_0} - N_{l_0}^{t_0} + \sum\limits_{i=1}^{n}(M_{r_i}^{t_i} - M_{r_{i-1}}^{t_i}) \\[2mm]
M_{r_0}^{t_0} - N_{l_0}^{t_0} + \sum\limits_{i=1}^{n}(M_{r_i}^{t_i} - M_{r_{i-1}}^{t_i}) + N_{r_n}^{t_n} - M_{r_n}^{t_n}
\end{cases}
\tag{7.7}
$$

In Eq. 7.7, $(M_{l_i}^{t_i}, M_{r_i}^{t_i})$ represents the $i$-th keypoint pair of hoop B at time $t_i$, while $(N_{l_i}^{t_i}, N_{r_i}^{t_i})$ denotes the corresponding keypoint pair for hoop A. Each formula corresponds to the respective cases illustrated in Figure 7.4.

Given that the iron chain moves with the machine during production, the coordinates of keypoints change over time. To address this issue, we compute relative distances instead of using absolute positions. This approach ensures consistent measurements despite the motion of the chain. Additionally, to enhance the connection between hoops across different time instances, we employ a tracking module as described by Zhang et al. [253]. This tracking method helps maintain the continuity and accuracy of keypoint associations over time.

## 7.4  EXPERIMENTS

We utilize a server equipped with the following computational resources: an NVIDIA GeForce V100 GPU, an Intel Core i7-8700K CPU, and 16GB of RAM. The system's performance is evaluated using a custom dataset captured by a pair of Intel SR305 cameras. This dataset includes 1000 RGBD images, each meticulously annotated with ground truth labels by factory workers.

For the RGB images, we annotated the endpoints of each hoop and recorded the manually measured lengths of individual hoops and the total chain lengths using rulers. To enhance the algorithm's generality and robustness, we diversified our dataset by including iron chains against various backgrounds and within different scenes.

### 7.4.1 Training Result on Chain Detection and Keypoints Detection

We allocate 80% of the images for training, 10% for validation, and the remaining 10% for testing. The batch size is set to 24. For learning rate scheduling, we adopt a step learning policy with a base learning rate of 0.001, gamma of 0.1, and step size of 800. The momentum is set to 0.9, and the total number of iterations is 10. The Adam optimizer is utilized for model training. For chain detection and keypoint extraction, we use pre-trained models: a YOLO model trained on the COCO dataset and a ResNet model trained on ImageNet, respectively.

Figure 7.5 shows some results of our own datasets. Figure 7.5(a)and 7.5(b) are the results of chain detection. Figure 7.5(c)and 7.5(d) showcase the estimated keypoints.

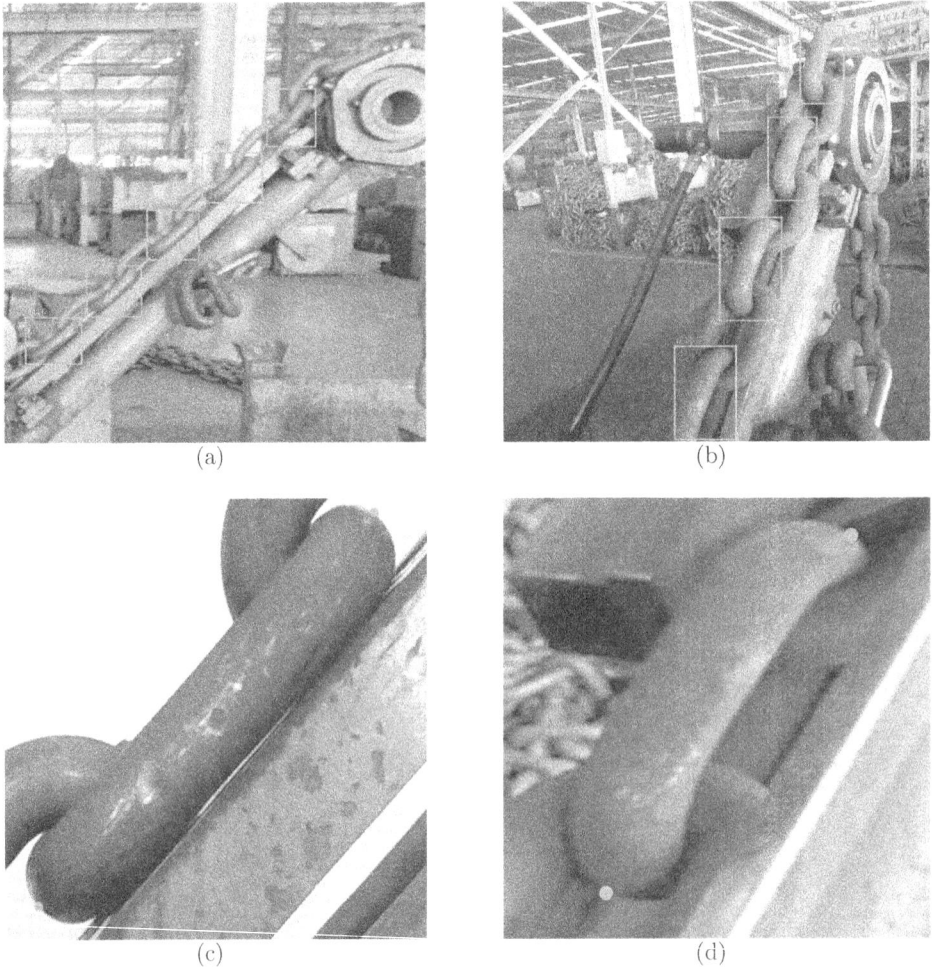

(a)

(b)

(c)

(d)

Figure 7.5   Detection results on the collected images.

As evidenced by these visualizations, our method achieves accurate detection and keypoint extraction on RGB images.

### 7.4.2 Comparison Experiment

To validate the effectiveness of our adopted models, we compared their performance against several other deep learning models. To ensure real-time computation capabilities, we deliberately avoided computationally intensive models such as EfficientDet [212] and transformer-based architectures like Swin Transformer [142]. Instead, we selected lightweight networks for comparative analysis, including Faster R-CNN [182], CenterNet [51], and YOLOv5 [98].

This approach allows us to evaluate the trade-offs between model performance and computational efficiency, ensuring that our solution remains practical for real-time industrial applications.

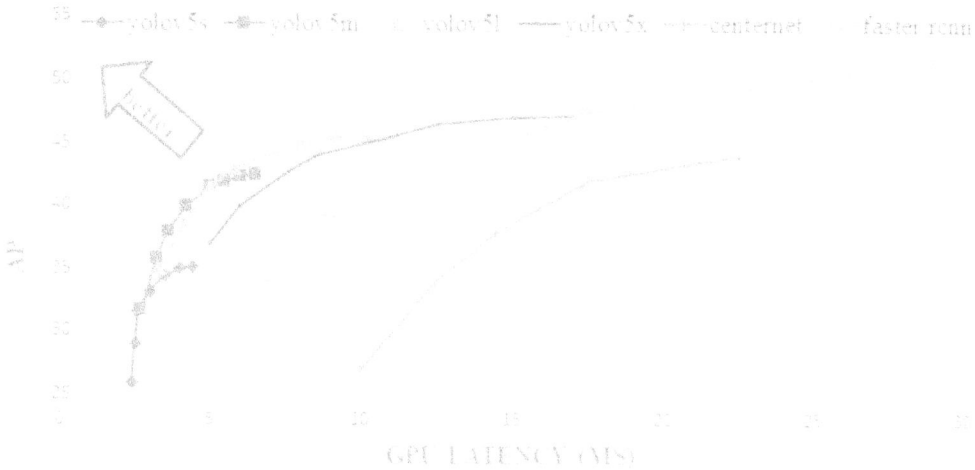

Figure 7.6  Comparisons of different detection methods on the collected dataset.

Figure 7.6 presents the quantitative evaluation results on our custom dataset. The X-axis denotes the running time, while the Y-axis represents the Average Precision (AP), a measure of detection accuracy. A higher value closer to the upper left corner indicates better performance. As shown, Faster R-CNN achieves the highest mean AP but has a relatively slower processing speed. Being a two-stage detector, it offers the best accuracy at the cost of computational efficiency. In contrast, the YOLO series models (e.g., [85, 98, 13]) strike a commendable balance between speed and accuracy. Specifically, YOLOv5l demonstrates an optimal trade-off for our application. Therefore, we selected the YOLOv5l model as the detector in our system.

Table 7.1 presents the performance metrics on test-dev data using keypoint estimation models trained on our custom dataset. We compared the faster ResNet-50 with the more accurate ResNet-101 to determine the most suitable backbone network. Additionally, we experimented with two different image crop sizes for the keypoint estimator. Based on the results in Table 7.1, we selected the ResNet-101 model with

TABLE 7.1   Results of the keypoints estimation module candidates.

| Keypoints Module | AP | AP.5 | AP.75 | AP(M) | AP(L) | AR | AR.5 | AR.75 | AR(M) | AR(L) |
|---|---|---|---|---|---|---|---|---|---|---|
| ResNet-50(256*192) | 0.651 | 0.863 | 0.732 | 0.618 | 0.705 | 0.703 | 0.89 | 0.771 | 0.661 | 0.768 |
| ResNet-50(384*256) | 0.663 | 0.861 | 0.730 | 0.641 | 0.721 | 0.721 | 0.895 | 0.772 | 0.675 | 0.787 |
| ResNet-101(256*192) | 0.658 | 0.862 | 0.741 | 0.645 | 0.710 | 0.714 | 0.896 | 0.776 | 0.664 | 0.784 |
| ResNet-101(384*256) | **0.685** | **0.873** | **0.753** | **0.660** | **0.736** | **0.738** | **0.902** | **0.798** | **0.682** | **0.804** |

TABLE 7.2   Measuring errors of the proposed method.

| | 1 | 2 | 3 | 4 | 5 | 6 | 7 | 8 | 9 | 10 | avg |
|---|---|---|---|---|---|---|---|---|---|---|---|
| our method | 25.6 | 24.3 | 24.5 | 25.8 | 25.4 | 48.9 | 49.2 | 51.9 | 52.2 | 48.3 | \ |
| ground truth | 25 | 25 | 25 | 25 | 25 | 50 | 50 | 50 | 50 | 50 | \ |
| diff | 2.40% | 2.80% | 2.00% | 3.20% | 1.60% | 2.20% | 1.60% | 3.80% | 4.40% | 3.40% | 2.74% |

TABLE 7.3　Comparisons with different methods.

| | accuracy | | time | |
|---|---|---|---|---|
| | single scene | multi scenes | single scene | multi scenes |
| manual method | 100% | 100% | 181s | 182s |
| traditional keypoints method | **98.50%** | 60.30% | **118s** | 125s |
| our method | 97.74% | **95.56%** | 121s | **122s** |

an image resolution of 384x256. This configuration offers a superior balance between accuracy and computational efficiency, making it ideal for our application.

The traditional keypoint detection method relies on conventional image processing algorithms, with gradient filtering being a common approach. In our experiments, we observed that chain hoops exhibit distinct gradient features, enabling the identification of each hoop's endpoints directly through gradient filters.

Table 7.3 shows that under controlled conditions, the traditional keypoint method achieves higher accuracy compared to our proposed method. However, as the scene complexity increases, the accuracy of the traditional method drops significantly, while our method demonstrates greater robustness and stability.

Regarding time consumption, Table 7.3 indicates that there is minimal difference in processing time between the traditional keypoint method and ours. In contrast, manual methods require substantially more time. These findings highlight the advantages of our method in terms of adaptability and efficiency in varying environments.

Figure 7.7 provides a detailed comparison between the traditional keypoint detection method and our proposed approach. When the background is simple, the blue star points generated by the traditional method closely match the red triangle points produced by our method. However, as the background complexity increases, the blue star points tend to drift or even disappear in some cases. This demonstrates that the traditional method is significantly more sensitive to background changes compared to our approach, which consistently shows better performance across various scenes.

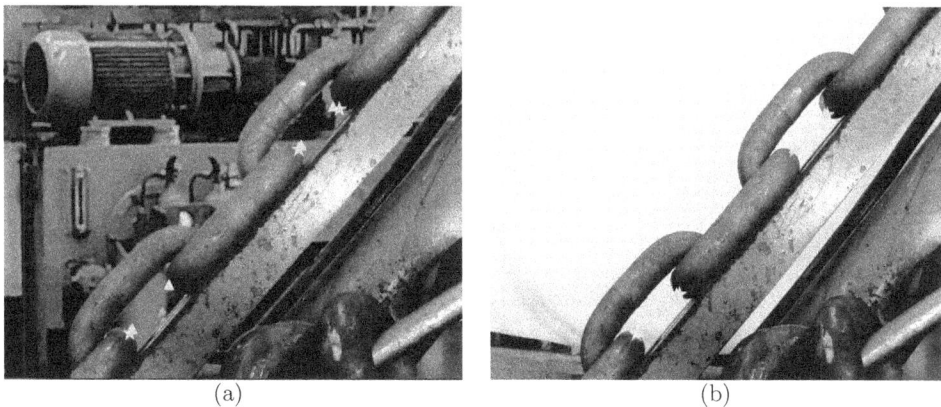

(a)　　　　　　　　　　　　　　　　(b)

Figure 7.7　Comparison of traditional keypoints method (star) and our method (triangle).

Table 7.2 compares our method against ground-truth data to evaluate the size estimation error for industrial items. As shown in the table, our method maintains an error rate within 3%, satisfying the stringent requirements of chain production.

## 7.5 CONCLUSIONS

This chapter presents an AI-driven object measurement framework, using iron chain length measurement as a case study to illustrate the entire process, which includes data acquisition, keypoint detection, and size calculation. Experimental results demonstrate that this approach achieves higher accuracy, faster processing speeds, and requires less expensive equipment. It also shows robustness to background variations, maintaining a measurement error rate consistently below 3%. Consequently, the proposed methodology effectively meets the demands of industrial production environments.

# CHAPTER 8

# Quality Estimation Paradigm for Copper Scrap Granules Recycling

## 8.1 INTRODUCTION

The scrap material recycling industry plays a crucial role in reducing the cost and environmental impact of producing new products [66]. One of the key steps in the recycling process is estimating the quality of scrap materials before they are recycled, as this helps determine their value and price [67]. In this context, quality estimation refers to predicting the content of valuable materials in scrap, specifically by calculating the mass proportion of valuable materials relative to the total mass of the scrap.

Currently, quality estimation relies on technicians who manually sample the scrap materials multiple times, stir them repeatedly, and identify various impurities based on their experience. However, this manual evaluation method suffers from several significant drawbacks. First, it is highly dependent on the experience and expertise of the technician, and training these experts is time-consuming and costly, resulting in extremely scarce expert resources. In addition, manual methods are time-consuming, laborious, and inefficient when dealing with large amounts of scrap. More importantly, the subjective nature of humans often causes manual methods to result in significant financial losses for the seller or purchaser.

With significant advancements in deep learning and computer vision [78, 151], it is now feasible to automatically estimate the quality of scrap materials. To overcome the limitations of manual quality estimation methods, this chapter introduces a general AI paradigm for automated quality estimation. In the following, we will first introduce the proposed quality estimation paradigm, which comprises image acquisition, image segmentation, and quality calculation. Then, we will apply this paradigm to the copper scrap recycling industry, demonstrating its effectiveness through comprehensive experiments and highlighting its broad range of applications.

DOI: 10.1201/9781003644972-8

## 8.2 QUALITY ESTIMATION PARADIGM

Our quality estimation paradigm involves three main steps: image acquisition, image segmentation, and quality calculation. The processing flow of this paradigm is illustrated in Figure 8.1.

### 8.2.1 Image Acquisition

For any type of scrap materials to be estimated, we first obtain a set of samples using a simple sampling method. The sample set is denoted as $S = \{s_1, s_2, ..., s_w\}$. Then, for each sample $s_i$, we stir it $n$ times and capture images after each stirring, obtaining the image set $c_i = \{c_1^i, c_2^i, ..., c_n^i\}$. The stirring method is not strictly defined but should ensure that valuable materials and impurities have an equal chance of being exposed on the top surface and captured by the camera. After sampling and capturing, we obtain the image sets $C = \{c_1, c_2, ..., c_w\}$, effectively transforming the quality estimation problem into an image recognition task.

### 8.2.2 Image Segmentation

Each image $c_j^i$ may contain various impurities in addition to valuable materials of interest. It is, however, impractical to use supervised image segmentation

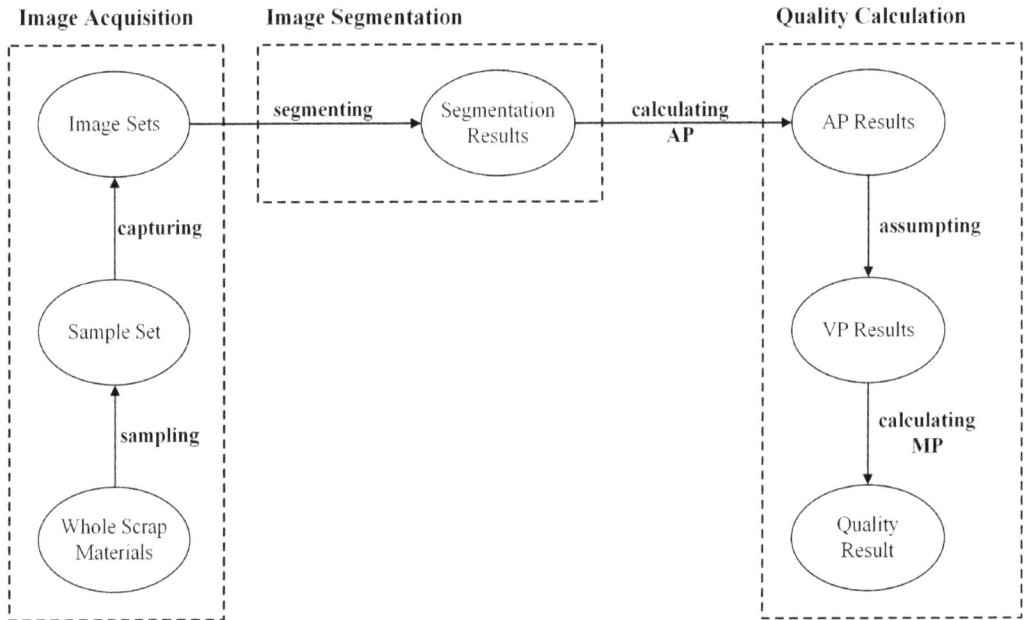

Figure 8.1 The process of quality estimation paradigm. AP represents the area proportion of valuable materials to the whole scrap materials in the image, and VP represents the volume proportion of the valuable materials to the whole scrap materials in the corresponding sample. MP represents the mass proportion of valuable materials to the total mass of the whole scrap materials, i.e., the final quality result.

networks to directly identify all potential impurities. One challenge is that training such a segmented network requires substantial annotation effort, complicating the paradigm's deployment in different application scenarios. Furthermore, it is impossible to anticipate all possible impurity types in many cases. To address these issues, we classify all types of impurities into a single category, and such simplified treatment contributes to the wide application of the paradigm. In our paradigm, for each input image $c_j^i$, the image segmentation network outputs a binary image $r_j^i$ where each pixel is accurately labeled as belonging to the valuable material region $R_V$ or another region $R_I$, that is, the impurity region. In this way, the segmentation network can effectively distinguish between valuable materials and others, simplifying the identification process.

### 8.2.3 Quality Calculation

#### 8.2.3.1 AP Calculation

Using the binary image $r_j^i$, we first calculate the proportion of pixels classified as $R_V$ to the total number of pixels in $c_j^i$. This proportion is defined as $a_j^i$, representing the AP. Then, for each scrap sample $s_i$, the set of AP for all $n$ images is denoted as $A_i = \{a_1^i, a_2^i, ..., a_n^i\}$. Finally, the average area proportion of valuable materials to the whole scrap materials (average AP) for sample $s_i$ is calculated as

$$\bar{a}_i = \frac{1}{n} \sum_{j=1}^{n} a_j^i \tag{8.1}$$

#### 8.2.3.2 Reasonable Assumption

Since the image acquisition process ensures that valuable materials and impurities are equally likely to appear on the top surface and be captured by the camera, it is reasonable to assume that the average area proportion of some stuff to the whole materials in the images is equivalent to its average volume proportion to the whole materials (average VP) in the sample. This assumption aligns with similar findings in previous work [248]. Therefore, the average VP for sample $s_i$ is denoted as $\bar{v}_i = \bar{a}_i$.

#### 8.2.3.3 MP Calculation

To calculate the mass proportion (MP) for sample $s_i$, we apply the mass calculation formula from physics

$$m_i = \frac{\rho \times \bar{v}_i \times V_i}{M_i} \tag{8.2}$$

Where $\rho$ is the density of the valuable material, $V_i$ and $M_i$ represent the volume and mass of the scrap sample, respectively, which can be measured directly using appropriate weighing tools. Finally, the final quality result for the entire input scrap sample set S is calculated as

$$\bar{m} = \frac{1}{w} \sum_{i=1}^{w} m_i \tag{8.3}$$

The proposed paradigm provides a systematic and general approach to estimating the quality of various scrap materials, improving the efficiency, objectivity, and accuracy of manual methods. It can be widely applied across various recycling scenarios, significantly enhancing the quality estimation process.

## 8.3 IMPLEMENTATION AND EXPERIMENTS

Copper is a valuable non-renewable resource, and copper scrap granules are a significant source of recycled copper often containing various impurities. Accurately estimating the quality of copper scrap granules holds significant economic and environmental value. In this section, we implement the proposed paradigm for the quality estimation of copper scrap granules to demonstrate its effectiveness. First, we construct a dataset comprising a training set, a validation set, and a test set. Next, we train a state-of-the-art image segmentation network and perform the quality estimation task according to the paradigm. Finally, we illustrate the validity of our paradigm through sufficient experiments.

### 8.3.1 Dataset

To train the segmentation network and evaluate our method, we create a practical dataset using recycled copper scrap granules of varying quality. Firstly, we collect 200 samples through simple sampling without replacement from a larger batch of copper scrap granules. Then, we stir each copper scrap granule sample 64 times and capture corresponding images. Finally, The dataset is then split into 160 samples for training, 20 for validation, and 20 for testing, totaling 10,240 training images, 1,280 validation images, and 1,280 test images. Some visual examples of copper scrap granules are shown in Figure 8.2.

When generating ground truth for each sample, we only label a quality scalar and 64 binary semantic segmentation images. The quality scalar for each sample $s_i$ is labeled by professional technicians specializing in copper scrap recycling. The area proportion (AP) for each image $c_j^i$ is then automatically computed based on its corresponding binary segmentation image $r_j^i$.

### 8.3.2 Implementation

We first construct and train a DeepLabv3+ [29] segmentation network using the train-val dataset to recognize copper and non-copper (impurities) regions, producing a binary segmentation result image $r_j^i$. DeepLabv3+ is a state-of-the-art deep learning model for image semantic segmentation, which is the process of partitioning an image into regions of interest, each labeled with a specific class. It features an encoder-decoder structure where the encoder extracts robust features from the input image using atrous convolution [29, 28], and the decoder refines and upsamples these features to produce high-resolution segmentation maps. The architecture of the copper scrap granule image segmentation network is illustrated in Figure 8.3.

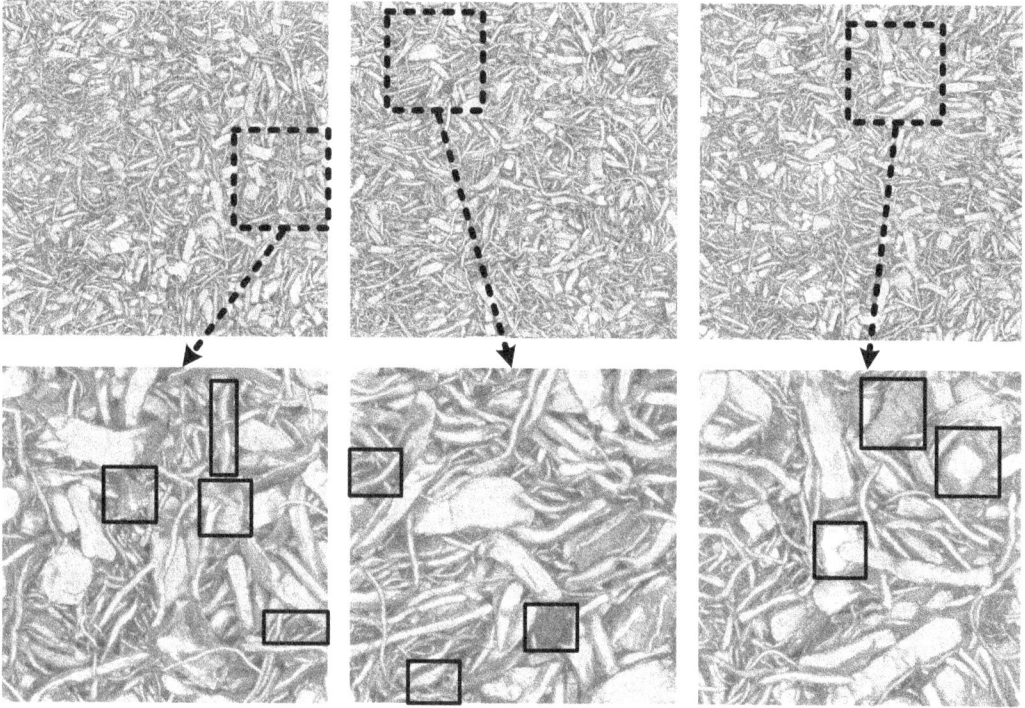

Figure 8.2   Images of copper scrap granules (top row) and their enlarged local views (bottom row). Rectangles highlight some impurities. Copper scrap granules typically contain various impurities with irregular shapes and different proportions, aside from copper.

During training, we optimize the network parameters using the Adam optimizer [261], and employ the cross-entropy loss function to measure the accuracy of the predicted segmentation results against the ground truth labels. The loss function is shown as follows:

$$Loss\,(\hat{y}, y) = -\,(1/p) \sum\nolimits_{i=1}^{p} y_i log \hat{y}_i \qquad (8.4)$$

Where $p$ is the number of pixels in the input image, $y_i$ represents the ground truth label of the $i_{th}$ pixel and $\hat{y}_i$ denotes the predicted probability of the $i_{th}$ pixel labeled as copper.

We perform regular validation using the 19,200 image validation set to continuously monitor the model's performance and adjust hyperparameters to prevent overfitting. After 100 epochs, the model improves and learns to segment various complex images accurately.

After training, we apply the proposed paradigm to the test samples, generating 19,200 segmentation result images and calculating 300 quality results for the entire test set.

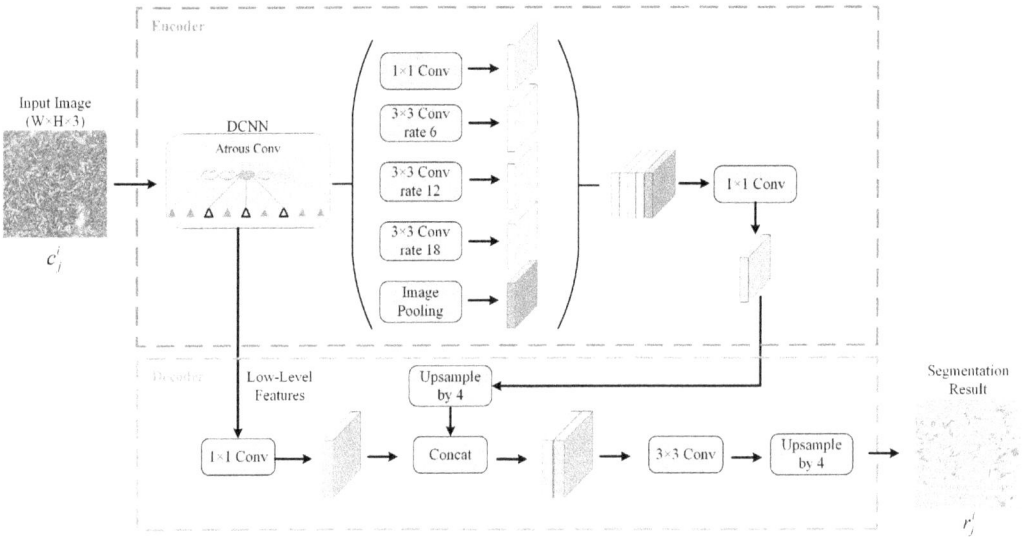

**Figure 8.3** Structure of the copper scrap granule segmentation network based on DeepLabv3+. The segmentation output is a binary image: dark regions indicate various types of impurities with irregular shapes (non-copper), while the remaining region represents copper.

### 8.3.3 Experimental Results

#### 8.3.3.1 Segmentation Results

We evaluate the segmentation network's performance using the mean Intersection over Union (mIoU) metric [151], a standard measure for segmentation accuracy. The final mIoU scores on the training, validation, and test sets are 0.904, 0.878 and 0.874, respectively. These results indicate that the segmentation network can segment the valuable (copper) regions from the impurities accurately, and has a strong ability to generalize from the training data to unseen data. It provides a solid foundation for subsequent quality estimation. Figure 8.4 illustrates examples of segmentation results.

#### 8.3.3.2 AP Results

Using the segmentation results, we calculate the AP results for each image in the test set, and the mean absolute error (MAE) across all test images is 3.67%, computed as:

$$MAE = \frac{1}{wn} \sum_{i=1}^{w} \sum_{j=1}^{n} \left| a_j^i - \hat{a}_j^i \right| \tag{8.5}$$

The AP result confirms the segmentation network's ability to accurately estimate the area proportion of copper to the total scrap material in various test images.

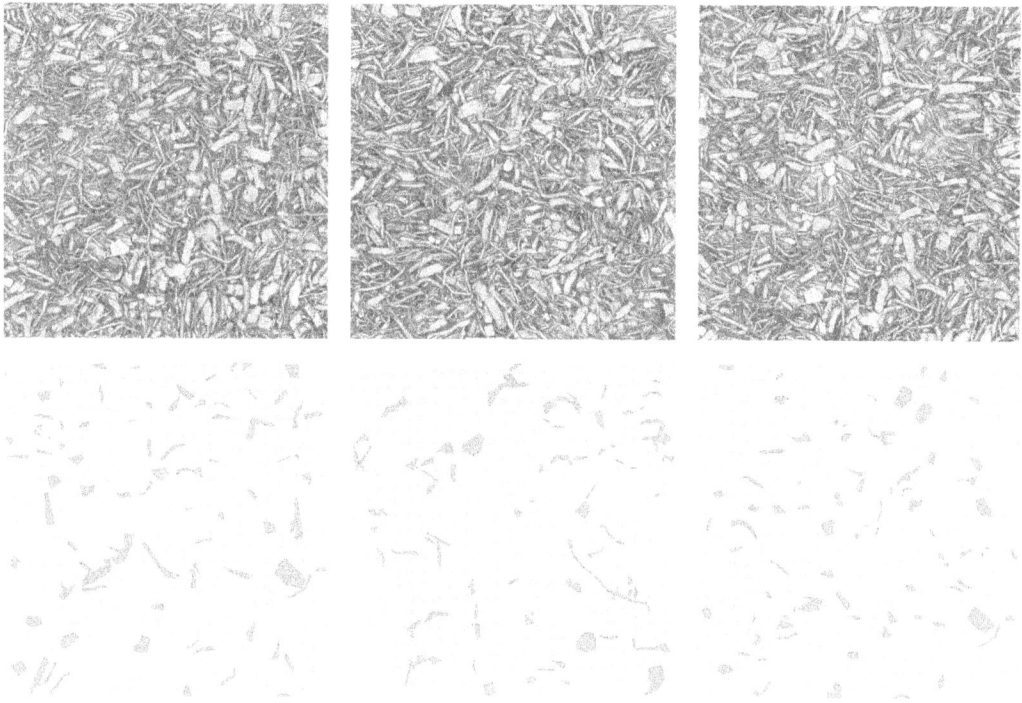

Figure 8.4 Examples of input (top row) and output (bottom row) images of the segmentation network. The dark regions represent impurities, while the remaining areas represent valuable material (copper).

### 8.3.3.3 Final Quality Results

We further calculate the quality of each copper scrap granule sample. The MAE for the quality estimation on the training, validation, and test sets are 2.04%, 3.13%, and 4.88%, respectively. Figure 8.5 illustrates the predicted results and corresponding ground truth for test samples.

The results demonstrate that our proposed quality estimation paradigm is effective and practical for estimating the quality of scrap materials in the recycling industry. Beyond copper scrap granules, the paradigm can be adapted for other recycling scenarios, such as steel, plastic, and wood scrap. Notably, when applying the paradigm to these other materials, the only additional effort required is collecting and labeling the relevant image data.

## 8.4 CONCLUSION

This chapter introduces a general quality estimation paradigm designed to address the limitations of traditional manual methods, including the shortage of experts, inefficiency, and subjectivity, while accurately evaluating the content of valuable materials in scrap. This paradigm consists of three key steps: image acquisition, image segmentation, and quality calculation. First, the image acquisition process ensures that valuable materials and impurities in the scrap are uniformly collected and transforms

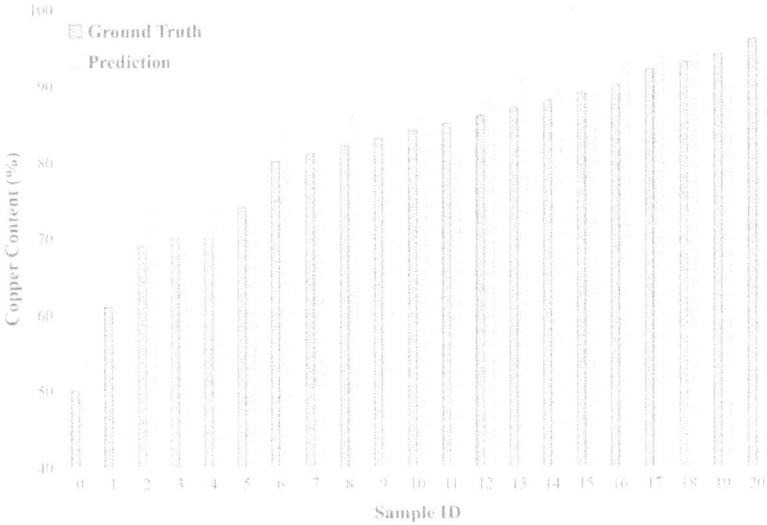

**Figure 8.5** Quality estimation results of our method compared to the corresponding ground truth on the test dataset. For convenience, the results are sorted in ascending order based on the ground truth.

the practical quality estimation task into an image recognition problem. Second, an image segmentation network is trained to accurately identify the regions containing valuable materials and impurities. Then, the area proportion (AP) is calculated based on the segmentation results. Finally, based on the reasonable assumption that the AP is equal to its volume proportion (VP), we design a direct and efficient method to calculate the quality result through physical principles. Using copper scrap granules recycling as a representative case study, we implement the proposed paradigm and demonstrate its effectiveness. Beyond copper scrap, the proposed paradigm can also be adapted to other recycling scenarios, such as steel, plastic, and wood scrap, showcasing its broad applicability.

# CHAPTER 9

# Human-in-the-Loop Learning Paradigm for Fabric Anomaly Detection

## 9.1 INTRODUCTION

In actual application scenarios, such as fabric production, steel production, and other industrial manufacturing fields, product anomaly detection (defect detection) is a key stage [76]. At present, product anomaly detection is usually achieved by manual labor, but there are two problems. On the one hand, the working environment is limited, the industrial production environment usually has high temperatures and loud noise, which makes it difficult to recruit suitable employees. On the other hand, work efficiency is limited, anomaly detection efficiency depends on the labor itself, and there is a relatively clear upper limit.

Currently, there are some industrial anomaly detection methods based on computer vision technology or artificial intelligence technology [214, 1, 103, 197, 129, 148, 59, 229, 245, 83, 236, 5, 82]. Computer vision algorithms (traditional or deep learning) are usually used to detect anomalies or defects in the collected industrial product images. However, these algorithms usually rely on enough abnormal/defective data samples, and when the data samples are insufficient, the algorithm can not ensure sufficient recognition rate performance. In practical application, the number of abnormal/defective samples is relatively small and it takes time to accumulate. This leads to the "chicken or egg" deadlock problem, which limits the rapid online application of the defect detection algorithm.

To solve the above problems, this chapter proposed a scheme, which consists of digital anomaly detection and intelligent anomaly detection. Digital detection refers to the digital upgrading and transformation of the physical anomaly detection system to realize manual digital anomaly detection and achieve the effect of "service is labeling." Intelligent anomaly detection is an upgrade of the digital anomaly detection system. It uses the data accumulated in digital anomaly detection to train the AI anomaly detection model and achieves the effect of replacing manual anomaly detection with AI anomaly detection. In this way, on the one hand, digital anomaly

DOI: 10.1201/9781003644972-9

detection can solve the problem of working environment limitation mentioned above (manual remote anomaly detection in air-conditioned rooms). Through the "service is labeling" process of digital anomaly detection, enough defective data samples can be accumulated to solve the pain points of insufficient data samples. On the other hand, intelligent anomaly detection can solve the work efficiency constraint (AI automated anomaly detection exceeds the limit of human efficiency).

The major contributions of this chapter: 1) Propose a new industrial anomaly detection scheme, which transfers from physical anomaly detection to digital anomaly detection and intelligent anomaly detection; 2) Realize and verify the effectiveness of the scheme by taking fabric detection as an example; 3) The scheme can be used as a general paradigm for industrial anomaly detection because it does not limit application scenarios and applies to a variety of low-speed anomaly detection scenarios.

The rest of the chapter is arranged as follows. Section 9.2 introduces the related work, including the existing industrial anomaly detection algorithm, the Human-in-the-loop idea, and the current physical fabric detection system, etc. Section 9.3 presents the proposed industrial anomaly detection scheme by taking fabric detection as an example. The scheme includes digital anomaly detection, digital-intelligent transition, and intelligent anomaly detection. Section 9.4 introduces the prototype of a digital fabric detection system and implements the AI fabric detection experiment to prove the feasibility of the scheme. Finally, Section 9.5 concludes the work of this chapter.

## 9.2 RELATED WORK

### 9.2.1 Industrial Anomaly Detection

At present, there are many industrial anomaly detection algorithms, which can be divided into several categories: algorithms based on image segmentation, algorithms based on object detection, and algorithms based on self-supervised/ unsupervised learning. In the algorithm based on image segmentation [148, 59], given the visual difference between the defective area and the normal area, image segmentation technology can be used to distinguish the defective area. This kind of method has a good segmentation effect on industrial objects with solid color, but the correct segmentation performance of industrial objects with complex textures will be significantly reduced. In the algorithm based on object detection [229, 245], the defective area is regarded as a special target, and the object detection algorithm based on deep learning is adopted to detect the defective target from the collected images. Such methods rely on sufficient anomaly data to train the object detection model and may be ineffective for new(previously uncollected) anomaly morphological types. For the algorithms based on self-supervised/unsupervised learning [83, 236, 5, 82], the self-supervised/unsupervised learning method is used to learn the texture/color pattern features of industrial items from the collected images, which are used as the feature representation of images or image blocks. Then, the abnormal areas can be determined by judging the similarity between the image feature representation of potential abnormal areas and that of normal areas. This method reduces the dependence on abnormal data samples and relies more on normal samples, but still requires a training and learning process. It can be seen that most algorithms need to be trained based on data samples, which requires solving the problem of data acquisition and accumulation. This chapter will build an iterative transition mechanism from manual detection to intelligent detection to achieve the acquisition and accumulation of required data samples and verify the iterative promotion process by taking the second and third algorithms as examples.

### 9.2.2  Human-in-the-Loop

Human-in-the-loop machine learning is to design some mechanism to make humans and machines (algorithms) interact and collaborate to better apply artificial intelligence technology to complete tasks [2, 257, 108]. Annotation and active learning are the basis of Human-in-the-loop. Labeling can add labels to the original data so that the original data can be processed into training data sets for AI model training. Active learning picks out the data most needed by the AI model and manually annotates it. It presents the data annotation process as an interaction between the learning algorithm and the user. For example, the Human-in-the-loop interactive data labeling framework trains the AI model through part of the data that has been labeled by users, marks the remaining data through this model, and then selects the data that is difficult for AI model labeling for manual labeling, and then uses these data for model optimization. After a few rounds, the AI models used for data labeling will have higher accuracy, allowing better data labeling and lower labeling costs. In this chapter, the idea of Human-in-the-loop will be used to make data annotation. The difference is that we build a "service is labeling" mechanism and integrate the "shadow" mechanism for training iteration and verification of the model.

### 9.2.3  Physical Fabric Detection System

The traditional physical fabric detection system usually consists of a fabric detection table, fabric guide roller, stepper motor, console and fabric inspector [113], as shown in Figure 9.1. The fabric detection table can spread the fabric flat and provide a light and length meter for the fabric inspector to easily check the condition of the fabric for anomaly detection. The fabric guide roller is connected to the stepper motor through the conveyor belt. When the stepper motor rotates, the fabric guide roller can be driven to rotate, so as to finally drive the fabric forward. The console provides a control button for the stepper motor, by which the fabric inspector can control the advance, retreat and stop the fabric. When the physical fabric detection system is used for anomaly detection, the fabric inspector controls the fabric movement through the console button, carries out a "naked eye" inspection of the cloth on the fabric detection table, and manually records the type, location and other data of anomaly, so as to complete anomaly detection. The physical fabric detection system needs to be inspected by the naked eye of the fabric inspector, which is easy to cause eye fatigue and visual loss, and the anomaly detection efficiency has an upper limit. Moreover, due to the high noise and high temperature of the industrial environment, it is difficult to recruit suitable workers. In this chapter, the physical fabric detection system is first digitally transformed, so that workers can cross the space limit to detect fabric

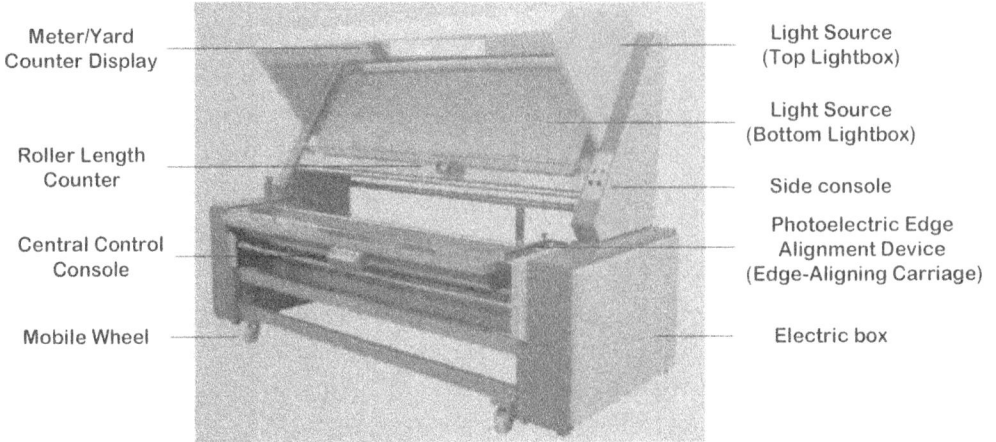

Meter/Yard Counter Display

Roller Length Counter

Central Control Console

Mobile Wheel

Light Source (Top Lightbox)

Light Source (Bottom Lightbox)

Side console

Photoelectric Edge Alignment Device (Edge-Aligning Carriage)

Electric box

(a)

Fabric Guide Roller

Fabric tension Regulator

External Fabric Stand

Fabric Basket

Rolling Tension Regulator

(b)

Figure 9.1   Physical anomaly detection machine.

in front of the computer screen. Then, the digital fabric detection is upgraded to an intelligent one, enabling AI to assist or replace workers in fabric detection to reduce manual participation and improve efficiency.

### 9.2.4   Visual Communication Technology

The rapid development of visual communication technology based on video compression technology and network communication technology makes real-time high-definition image/video communication collaboration possible. The new generation of video coding standard H.265/HEVC [233] has further improved video compression efficiency. Compared with H.264/AVC, the compression efficiency has increased by $30\% \sim 50\%$, and the performance improvement of 4K and 8K resolution ultra-high-definition video is more outstanding. The 5th generation mobile communication technology (5G) [246] has the characteristics of high speed, low delay and large connection.

The user experience rate can reach 1Gb/s, the delay can be as low as 1 ms, and the user connection capacity can reach 1 million connections/km2. The combination of H.264/HEVC video coding technology and 5G provides strong technical support for ultra-high-definition and low-delay real-time video collaboration [27]. In this chapter, the traditional physical fabric detection system is transformed into a digital one by using visual communication technology. Based on the Human-in-the-loop method, a human-computer collaborative closed-loop is formed, data is accumulated in the digital fabric detection to train the industrial anomaly detection algorithm to replace the digital fabric defect detection, and the intelligent upgrading and transformation of industrial anomaly detection is finally realized through iterative and progressive improvement.

## 9.3 PROPOSED AI PARADIGM FOR INDUSTRIAL ANOMALY DETECTION WITH HUMAN-IN-THE-LOOP

In the chapter, a common multistage anomaly detection solution is proposed. First, the local physical anomaly detection machine is upgraded to digital anomaly detection to support remote anomaly detection. Then the worker can detect anomalies remotely with the digital system, and the detected results will be saved during the process. Meanwhile, the accumulated data will be utilized to iteratively train the AI model and continuously improve its performance, in order to replace human quality inspection finally. As shown in Figure 9.2, the multistage solution consists of the following stages, such as local human anomaly detection stage, digital anomaly detection stage, digital-AI transition stage and AI-based intelligent anomaly detection stage. Among the multiple stages, local human anomaly detection indicates the traditional quality inspection where humans detect anomalies on the physical anomaly detection machine. In the digital anomaly detection stage, the physical quality inspection machine is upgraded, and then the human detects anomalies by manipulating the machine remotely. In the digital-AI transition stage, remote digital quality inspection is

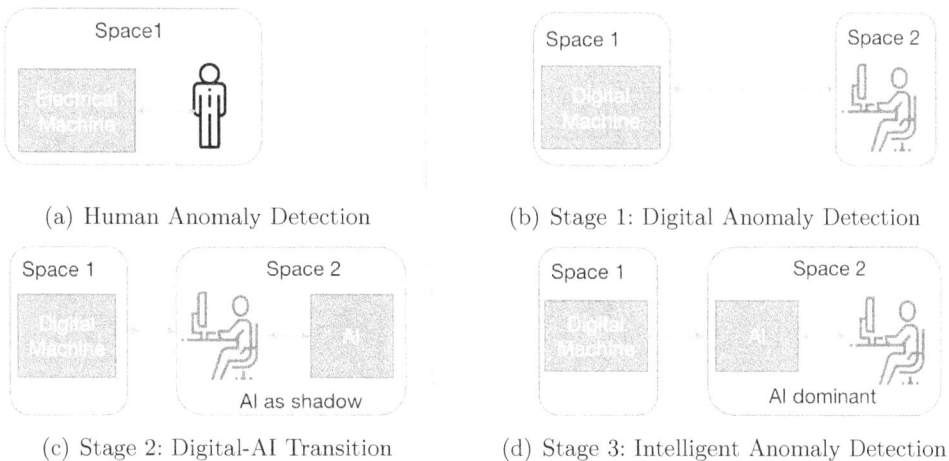

(a) Human Anomaly Detection      (b) Stage 1: Digital Anomaly Detection

(c) Stage 2: Digital-AI Transition      (d) Stage 3: Intelligent Anomaly Detection

Figure 9.2   Proposed anomaly detection paradigm.

also applied. However, the results of the quality inspection are continuously accumulated and applied to train and improve the performance of the AI model iteratively. Meanwhile, the trained AI model is running as a shadow to evaluate the gap between the performance of the AI model and that of humans. When the performance of the AI model reaches that of the human, the AI model will be brought to the foreground and do quality inspection by replacing human. So the AI-based intelligent anomaly detection is reached. In the following parts, the three stages except for the human quality inspection stage will be introduced in detail.

### 9.3.1 Digital Anomaly Detection

In the digital anomaly detection stage, the physical quality inspection machine is upgraded and digitalized to support remote and online anomaly detection. In order to present conveniently, fabric anomaly detection is adopted as an example in this chapter, and the solution can also be applied in other similar scenarios. The physical anomaly detection and the digital anomaly detection are shown in Figure 9.3. As shown in Figure 9.3(a), in the physical anomaly detection scenario, the worker and the machine are in the same working space, and the quality inspection is performed by the worker who manipulates the machine. The worker can control the motor run or stop by pushing the buttons on the machine and do a quality inspection by checking whether there are defects in the fabric with his eyes. If a defect is found, the machine will be stopped and the defect will be marked when the defect is confirmed. In the proposed digital anomaly detection solution, the physical anomaly detection machine

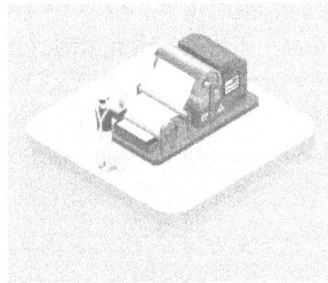

(a) Physical Anomaly Detection Scenario

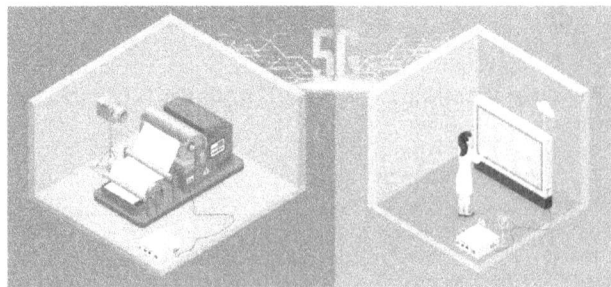

(b) Digital Anomaly Detection Scenario

Figure 9.3  Physical and digital anomaly detection scenarios.

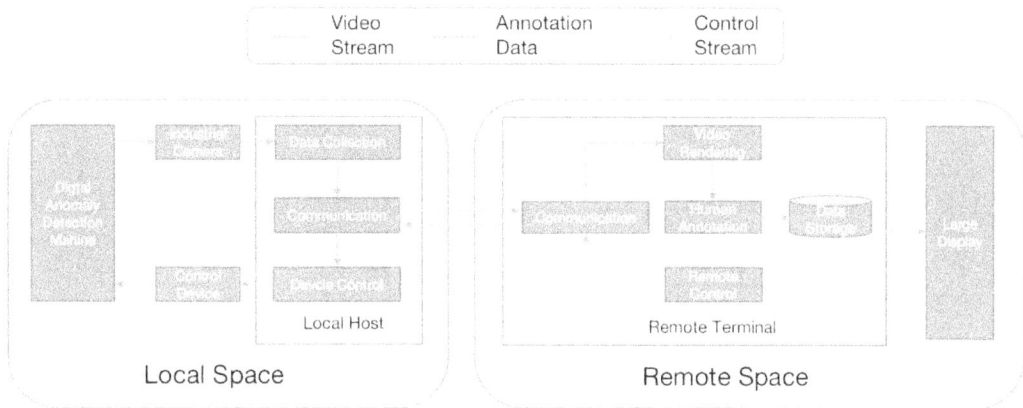

**Figure 9.4** Digital anomaly detection system architecture.

is upgraded to a digital anomaly detection machine by being connected to some digital devices, such as an industry camera, a motor controller, a host, a remote terminal, a large screen display, and so on. The system of digital anomaly detection is composed of two parts, the local digital anomaly detection machine and the remote terminal which are connected by high-speed networks such as 5G. Among the devices, a digital anomaly detection machine, an industry camera, a motor controller and a host are placed in one space, and a remote terminal and large screen are placed in the other space.

The working process in the digital anomaly detection stage is shown in Figure 9.4. During the process of quality inspection, the video of the fabric is captured by an industrial camera, and the video is compressed and sent to a remote terminal by a communication module. When the video is received by the remote terminal, it is uncompressed and presented on large display. The worker detects anomalies by watching the video played on a large display. When a defect is found, the worker will control the quality inspection machine move ahead or move back, and then will mark and record the defect. Once the detected anomaly is handled, the quality inspection machine will be controlled to resume. In the process mentioned previously, when the buttons on a large display are pressed, the terminal will send corresponding control commands to the host located at a remote space, and then the host will control the motor accordingly. Additionally, the remote terminal is responsible for saving the defect data found in the quality inspection process.

As presented in Table 9.1, the annotated defect data contains the information about the quality inspection task, the item under quality inspection, the annotation and related images. In detail, the information about the quality inspection task includes time, location, method, worker and so on. The information about the item under quality inspection includes the item's NO., type, description, details, etc. Annotation information includes the defect's position, size, type, etc. And the images have types, including original images and annotated images. An original image and an annotated image are shown in Figure 9.5(a) and Figure 9.5(b), respectively.

TABLE 9.1   An example of annotation data.

| No | Annotation Data Type | Annotation Data Valuet |
|---|---|---|
| 1 | Time | 2022-11-15 16:32:45 |
| 2 | Location | Build 2, Room 203 |
| 3 | Method | Online |
| 4 | Worker | John |
| 5 | Item NO. | F202211150004 |
| 6 | Item Type | Fabric |
| 7 | Item Description | Pure Black Fabric |
| 8 | Item Size | Width: 150cm, Height: 2000cm |
| 9 | Defect Position | X: 16.5cm, Y: 268cm |
| 10 | Defect Size | Width: 5cm, Height: 2cm |
| 11 | Defect Type | Hole |
| 12 | Original Image | Shown in 5(a) |
| 13 | Annotated Image | Shown in 5(b) |
| 14 | Defect Position In Image | X:2635, Y: 162 |
| 15 | Defect Size In Image | Width:49, Height:20 |

### 9.3.2   Digital-AI Anomaly Detection Transition

Based on the results of the digital anomaly detection stage, AI-model based quality inspection modules are added to build a digital-AI anomaly detection transition system. In the system, the annotated defect data accumulated in the digital anomaly detection stage are utilized to train the AI model iteratively. Generally, the AI model could be an object detection model (such as YOLOv6 [123]) or a self-supervised learning model Cutpaste [122]. As demonstrated in Figure 9.6, compared with a digital anomaly detection system, a digital-AI quality inspection transition system includes additional function blocks, including defect data accumulation and trigger, AI model training, AI model deployment, and AI model human-like performance evaluation. When the volume of defects data accumulated in the previous stage reaches a threshold, AI model training will be launched and the generated AI model will be deployed as an AI inference service. Then the AI inference service is running in the background to do quality inspection at the same time the human is working. The two results from the AI model and human are compared to evaluate the AI model's performance. The process normally iterates multiple times to improve the performance of the AI model progressively. In the whole stage, when a human detects an anomaly, the AI inference service is running as a shadow and does quality inspection simultaneously, but does not take part in the quality inspection procedure and does not influence the final result of quality inspection. So the stage can be regarded as "Human-Dominant and AI-assisted" anomaly detection.

Concretely, the working process of the digital-AI anomaly detection transition system is detailed below:

(a) Original Image

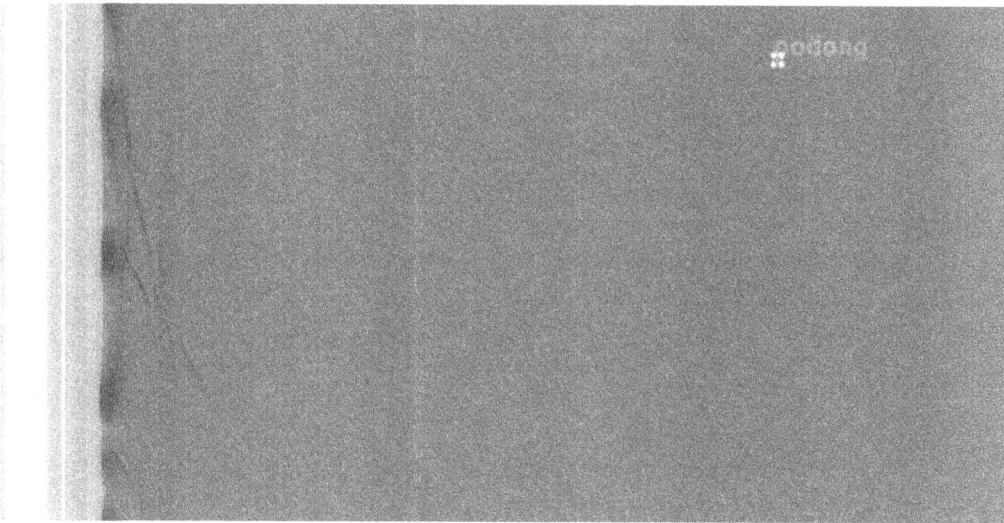

(b) Annotated Image

Figure 9.5    An example of original image and annotated image with defect.

Step 1: The defect data are collected during the digital quality inspection process continuously in a real-time manner. When the volume of defect data reaches a specified threshold (100 images as examples), a defect training data set, indicated by $S_i$, is prepared for the $I_{th}$ iteration training and then goes to step 2.

Step 2: the training data set prepared in Step 1, $S_i$, is utilized to train the pre-designed initial version of AI model $M_0$ or the AI model trained in the $(i-1)_{th}$ iteration $M_{i-1}$ and generate the $I_{th}$ version of AI model $M_i$. Then the generated AI model $M_i$ will be deployed as an inference service running in the background. The training process can be expressed as $M_i = T(M_{i-1}, S_i)$, where $T()$ means the training process.

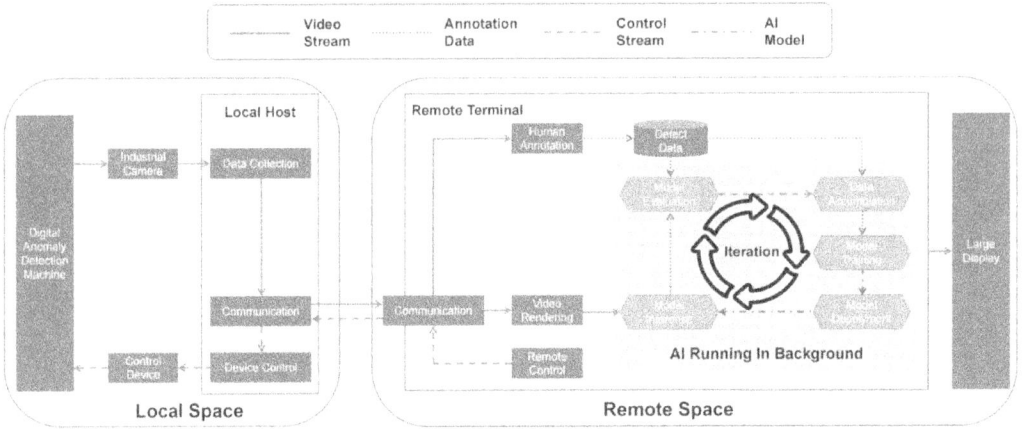

Figure 9.6   Digital-AI anomaly detection transition system architecture.

Step 3: During the human digital quality inspection, the inference service with AI model $M_i$ is also applied to real-timely detect defects from the fabric images simultaneously. The two results of AI and humans are also compared in order to evaluate the performance of the AI model. The gap between the performance of the two results is smaller, the performance of AI is better. Generally, a few images are exploited in the evaluation rather than a few images to get a more convincing evaluation result.

Step 4: If the evaluation result from Step 3 shows that the gap between the two results of AI and humans is still bigger than a specified threshold $D$, it means that the performance of the AI model has not reached the expected target. So go to Step 1 to continue accumulating data and improve the AI model, otherwise go to Step 5.

Step 5: Finish the transition stage and move ahead to the AI anomaly detection phase which will be introduced later.

### 9.3.3   AI-based Intelligent Anomaly Detection

The system architecture in the intelligent quality inspection stage is presented in Figure 9.7. Compared to the digital-AI transition stage, there are two differences. First, AI quality inspection is brought to the foreground from the background. The results of AI quality inspection are regarded as valid, and human anomaly detection becomes optional and will be triggered only when necessary. Second, a new module, named the human review module, is added for humans to review and check the results of AI anomaly detection and form an evaluation report. If the evaluation report shows that the performance of the AI model cannot meet the requirements, the human anomaly detection module will be reactivated and the whole evaluation process of quality inspection will return to the AI-human anomaly detection transition stage to improve the performance of the AI model again. When a human review is performed, the reviewer will check the images with defects detected by the AI model one by one in offline mode to confirm the rightness of the result. According to the review result, a review report will be generated and a metric R, named comprehensive accuracy rate,

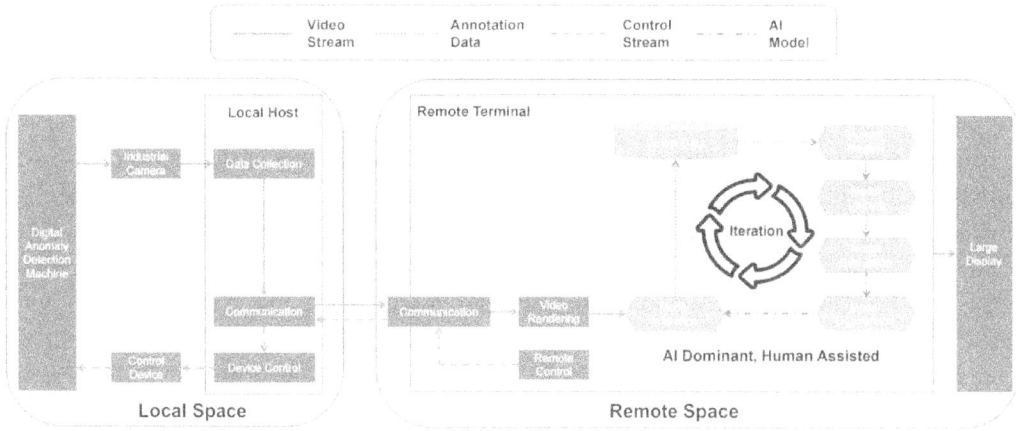

Figure 9.7    AI intelligent anomaly detection system architecture.

will be calculated by the statistics of the reviewed images. If R is less than 1-D, the human quality inspection module will be activated and the whole evolution process will return to the AI-human transition stage. The results of AI Anomaly Detection and Human Review are shown in Figure 9.8.

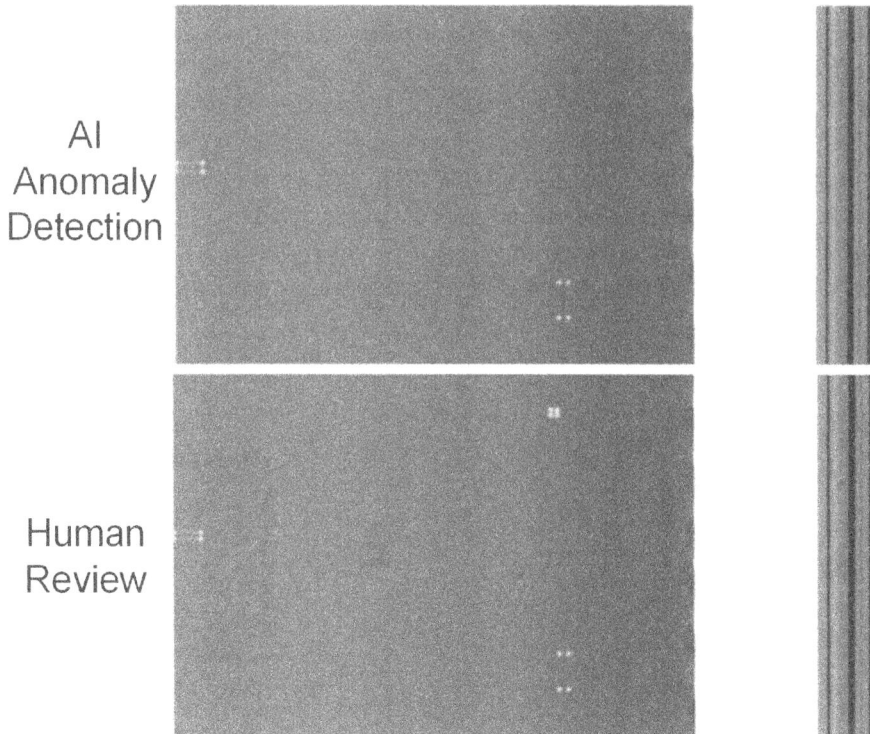

Figure 9.8    The results of AI anomaly detection and human review.

## 9.4 EXPERIMENTS

### 9.4.1 Prototype of Digital Fabric Inspection System

In order to verify the scheme proposed in this chapter, we built a prototype of the digital fabric inspection system based on the traditional fabric inspection machine, as shown in Figure 9.9. By adding industrial cameras and local hosts, and connecting the local host and the motor control system of the fabric inspection machine, the physical fabric inspection machine is transformed into a digital fabric inspection machine. The remote cloth inspection terminal is built by a NUC, which is connected to a display with a touch screen, and the local host and the remote terminal are connected through 5G CPE. Through H.265/HEVC video encoding and 5G network transmission, it can achieve 4K picture resolution and less than 50 ms transmission delay to meet the real-time quality inspection requirements.

On the remote fabric inspection system, the manual fabric inspection operating interface is shown in Figure 9.10. In the first step, the fabric inspectors open the fabric inspection system interface on the remote fabric inspection terminal, and select the console for fabric inspection, and the operating console of the fabric inspection machine will appear in the middle of the screen, and the previous fabric inspection data will appear on the right side of the screen. In the second step, the fabric inspector clicks the "Start" button on the lower part of the fabric inspecting system on the screen to control the start of the remote digital fabric inspecting machine and can see the real-time image of the remote fabric in the center of the screen to start the fabric inspection. In the third step, when a defect is found, the staff can click the "Pause" button to control the remote digital fabric inspection machine to stop, and then click the "annotation" button to enter the annotation interface to label the defect. After annotation is completed, the quality inspection data will be recorded. Save the inspection data and return to the second step to continue the fabric inspection. In the fourth step, when the fabric inspection is over, the fabric inspectors can click the "Stop" button to control the digital fabric inspection machine to stop. After

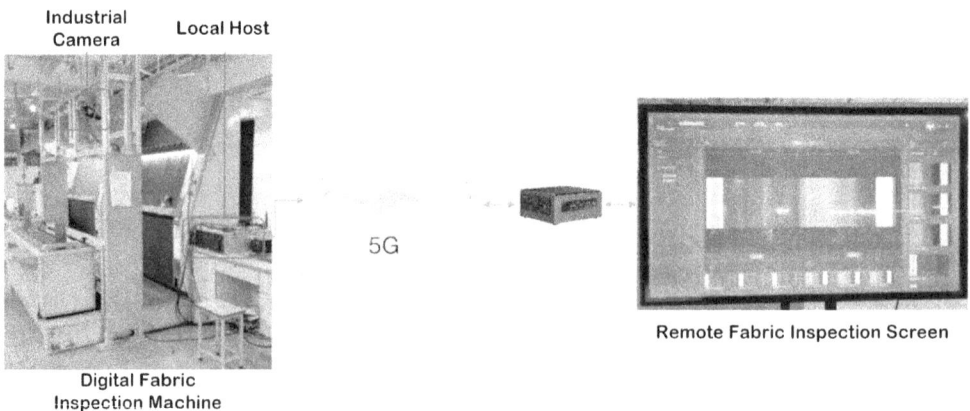

Figure 9.9 Prototype of digital fabric inspection system.

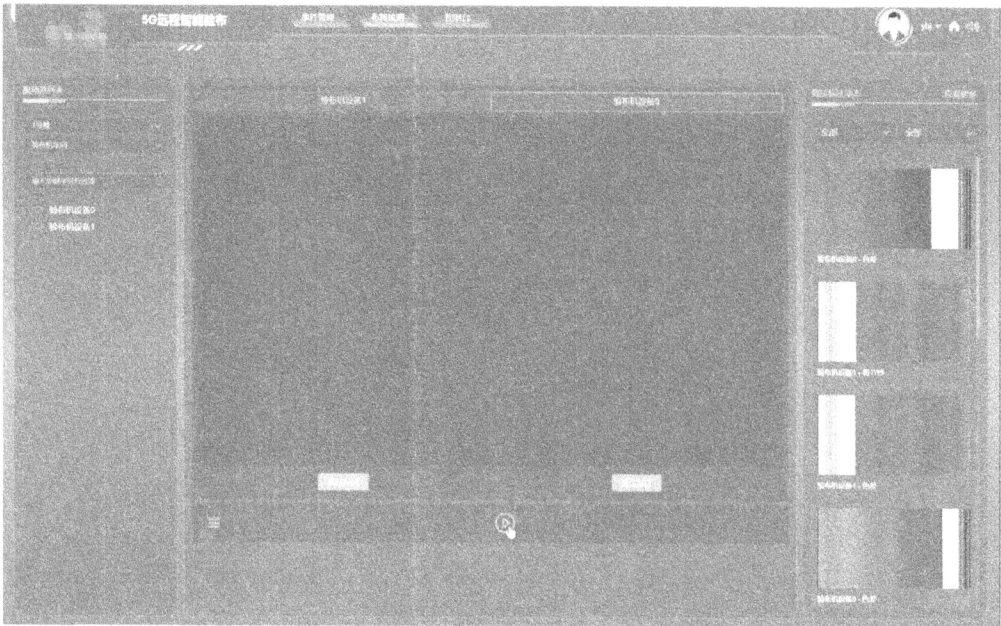

(a) Open the fabric inspection system interface

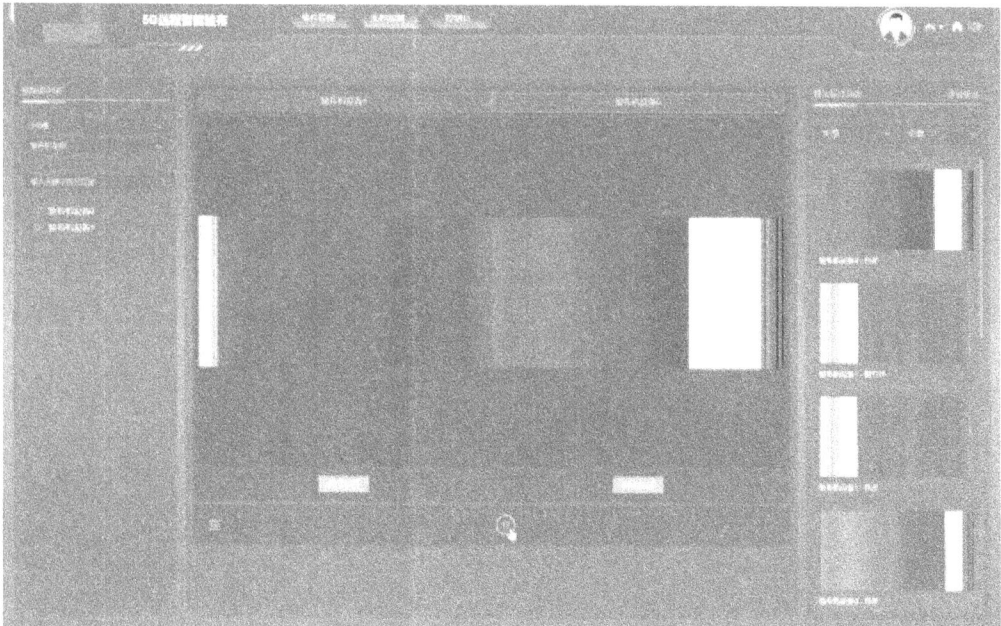

(b) Start the fabric inspecting machine

Figure 9.10  Manual fabric inspection operating interface of the remote fabric inspection terminal.

(c) Manual labeling operation

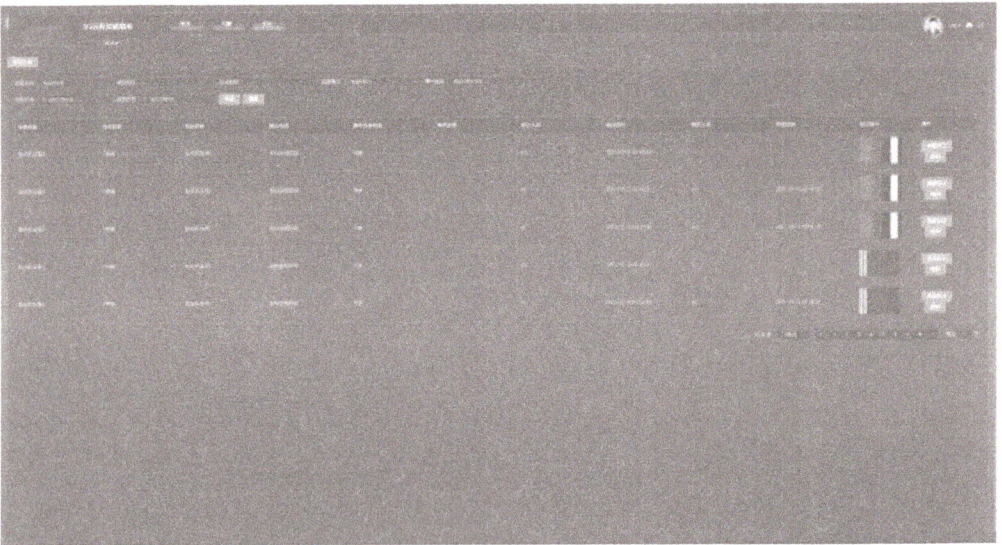

(d) quality inspection data query and management

Figure 9.10 (*Continued*).

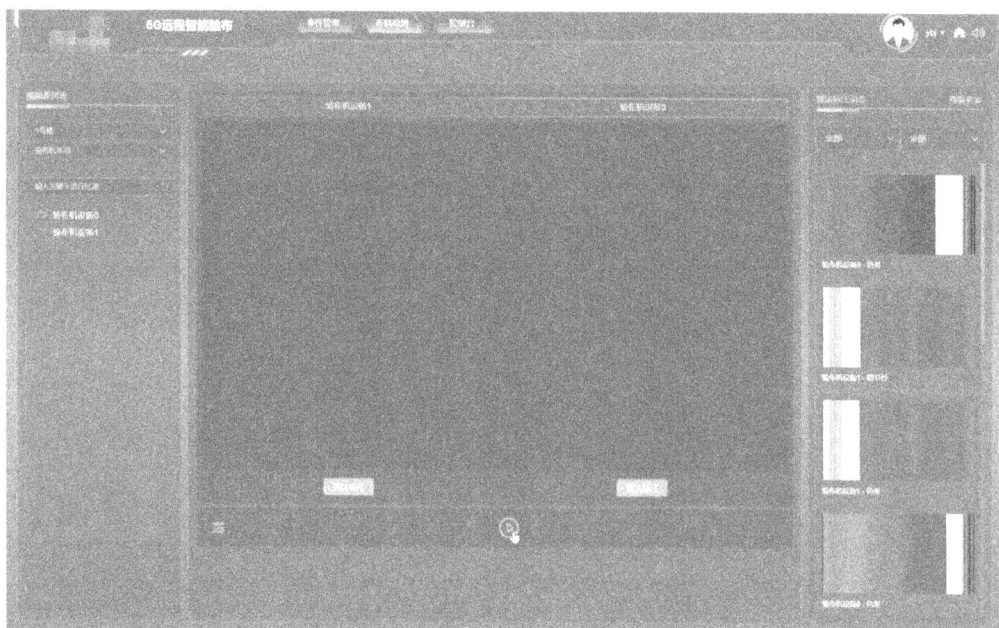

(a) AI quality inspection startup interface

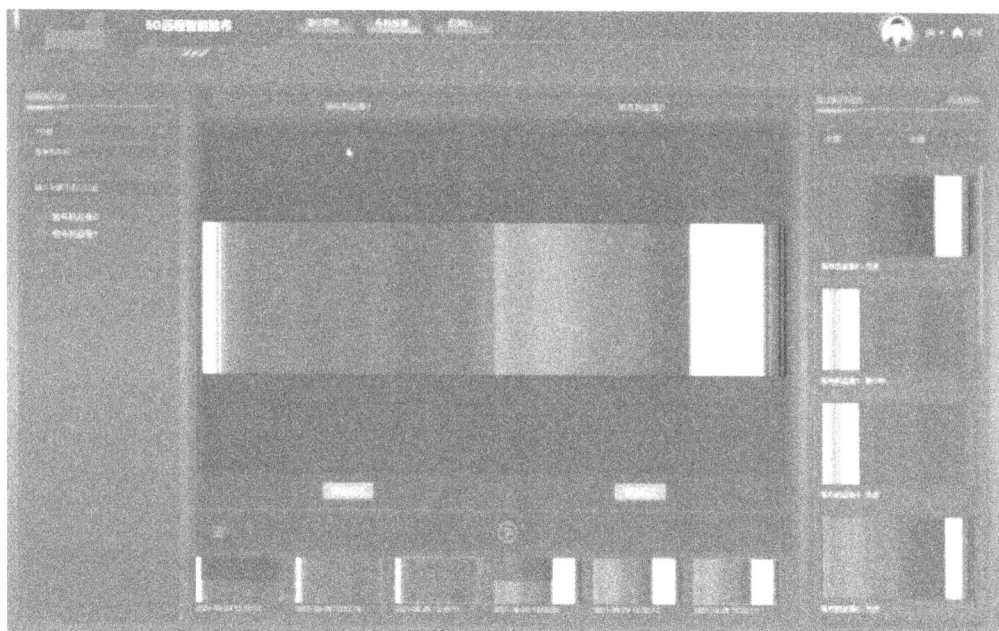

(b) AI quality inspection in progress

Figure 9.11　The intelligent fabric inspection operation interface of the remote fabric inspection terminal.

the fabric inspection is over, the fabric inspection personnel can view the quality inspection data generated during the fabric inspection process and management.

The current implemented intelligent fabric inspection operation process is shown in Figure 9.11. The steps of digital fabric inspection are basically the same. The

(c) AI quality inspection result query

(d) Manual review operation

Figure 9.11 *(Continued)*.

TABLE 9.2   Comparison of multiple quality inspection methods.

| function | quality inspection methods | | |
|---|---|---|---|
| | Local physical | Digital | Intelligent |
| Fabric Movement Control | Manual | Manual | AI |
| Defect labeling record | Manual | Manual | AI |
| Manual periodic review | Need | Need | Need |
| Environment restrictions | Yes | No | No |
| Work efficiency limits | Yes | Yes | No |

difference is that after the fabric inspection is started, defects will automatically detected by the AI quality inspection model. Anomaly detectors only need to review the results generated by the AI model.

We compared the functions of the implemented digital quality inspection and intelligent quality inspection with the traditional local physical quality inspection, as shown in Table 9.2. Compared with traditional local physical quality inspection, digital quality inspection can be carried out remotely online without being limited to the production site, thereby reducing the restrictions on the working environment; while intelligent quality inspection uses AI-driven machines to control fabric movement and automatically mark defects, without having to rely on manual quality inspection operations, thereby reducing the dependence on manual work efficiency without being limited.

## 9.4.2   AI Fabric Inspection Experiment

Experimental settings: There are 9 types of fabric texture, and 10 types of defects: seams, variegated wool, stains, watermarks, pleats, wrong texture, holes, color difference, thick warp, and thick weft.

Verification experiments have been done for two types of AI quality inspection models, namely the object detection model YOLOv6 [123] and the self-supervised learning model Cutpaste [122]. When the manually labeled data increases by 100%, the model is iteratively updated. The experimental results of manual and AI comparison are shown in Figure 9.12. The abscissa indicates the amount of image data, and the ordinate indicates the accuracy of AI recognition. Assume that the accuracy rate of manual quality inspection is 100%.

It can be seen the matching degree with manual quality inspection is getting higher while iterative training the same type of fabric (color or texture) model. From the result curve, it can be seen that the AI performance has less room for improvement. When the data volume is 700 When the matching rate reaches more than 95%, AI can be triggered to replace manual quality inspection operations.

In addition, the experiment verification of two different types of fabrics (color or texture) is shown in Figure 9.13, where the object detection model is taken as an example. Using the model that has been iteratively trained for fabric 1, the recognition rate of fabric 1 can reach 98% while using this model to identify fabric 2, the

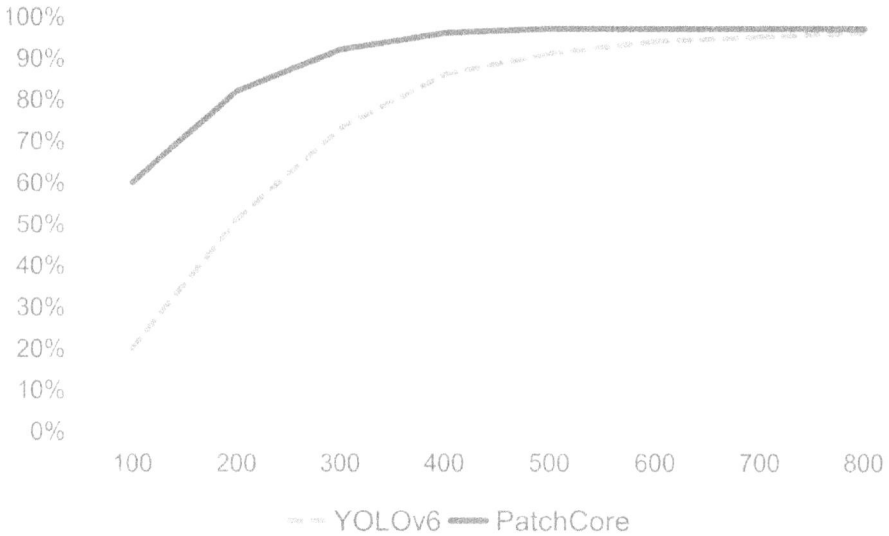

Figure 9.12   AI model iterative improvement process.

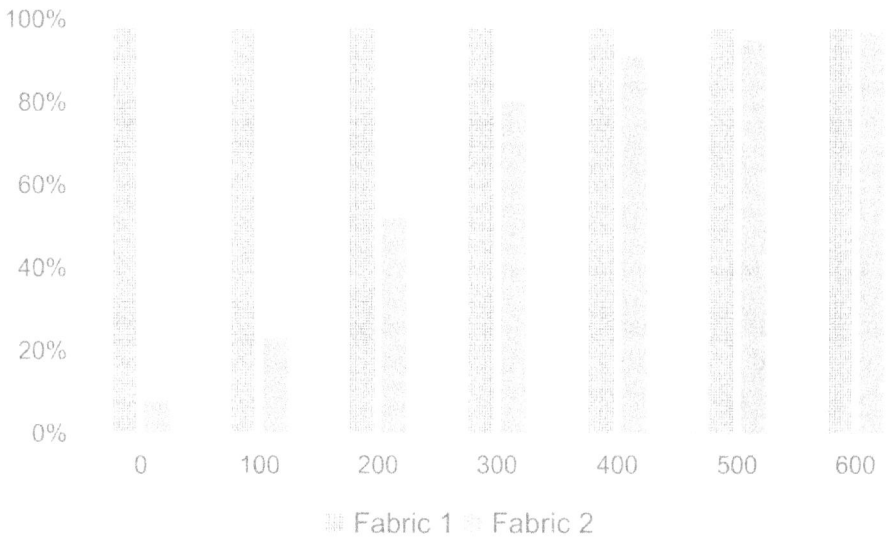

Figure 9.13   Iterative improvement process of AI model for different fabrics.

recognition rate is only 9%; but after collecting data for re-iteration, when the sample dataset size reaches 600, the recognition rate of fabric 2 continuously increased to 97%, and the recognition rate of fabric 1 remains unchanged at 98%.

## 9.5   CONCLUSION

This chapter proposes a general paradigm for solving industrial quality inspection problems with digitization and intelligence, which consists of three steps: digital quality inspection, digital-AI quality inspection transition, and intelligent quality inspection. Among them, in digital quality inspection, manual data annotation is done while doing quality inspection; in intelligent quality inspection, the system automatically uses the labeled data accumulated in the digital quality inspection process to train the AI quality inspection model and replace manual quality inspection; From digital quality inspection to intelligent quality inspection requires a transition process, that is, only the AI quality inspection model can be officially launched after passing the comparative evaluation with manual quality inspection, otherwise it can only be run as a "shadow" in the background. This comparative evaluation and replacement process can all be automated without manual triggering. Taking fabric quality inspection as an example, we implemented the prototype of digital fabric quality inspection and intelligent fabric quality inspection and adopted a typical AI quality inspection model to verify the effectiveness of iterative improvement of the AI fabric quality inspection model assisted by manual supervision. The scope of application of this paradigm is not limited to fabric quality inspection scenarios but also applies to other industrial quality inspection scenarios that run at low speed, such as engine quality inspection, vehicle body quality inspection, steel pipe quality inspection, etc., Therefore, this paradigm can be used as a general paradigm for intelligentization of industrial quality inspection.

# CHAPTER 10

# Supervised Learning Paradigm for Edible Oil Anomaly Detection

## 10.1 INTRODUCTION

In the industry, ensuring product quality is paramount, as defects can lead to substantial financial losses and damage to brand reputation. Traditional manual inspection methods for detecting defects in products, such as electronics product quality inspection, metal material quality inspection, bearing quality inspection, glass cover plate quality inspection, etc., are labor-intensive, time-consuming, and susceptible to human error. Inspectors must meticulously examine large quantities of products to identify defects, a process that can be inconsistent and inefficient due to human variability.

To address these limitations, this book presents a supervised industrial quality inspection paradigm, which leverages cameras to capture images from the production line. These images are then processed using visual detection models, so as to automate and intelligentize the quality inspection workflow.

Our paradigm is shown in Figure 10.1, which involves four main steps: data acquisition, image pre-processing, defect detection, and evaluation:

**Data Acquisition.** Data acquisition step involves collecting a large set of images of the products from the industrial production line. High-quality and diverse datasets are crucial for training a robust defect detection model. The data acquisition process should capture various possible defects under different lighting and environmental conditions to ensure the model generalizes well to real-world scenarios.

**Image Pre-processing.** Pre-processing the collected images is essential to improve the performance of the defect detection model. This step includes labeling images with annotations that indicate where defects are located, operations like resizing images to a standard size, normalizing pixel values, and augmenting the dataset by adding variations such as rotations, translations, flips, etc. Pre-processing helps in standardizing the input data and enhancing the model's ability to detect defects under different conditions.

DOI: 10.1201/9781003644972-10

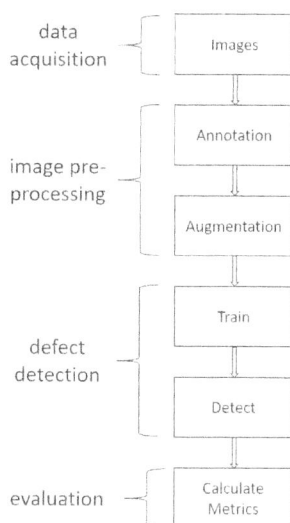

Figure 10.1    Workflow of our paradigm.

**Defect Detection.** In defect detection step, the pre-processed images are fed into the model, which identifies and localizes defects in real-time. It's the core task of the paradigm, where the model learns to distinguish between normal and defective product characteristics based on the training data, which is crucial for maintaining production efficiency and quality control.

**Evaluation.** After detecting defects, the evaluation step evaluates the performance of the model. Metrics are used to assess how well the model identifies defects compared to ground truth annotations. Evaluation helps in refining the model, improving its accuracy over time, and ensuring reliable quality control in industrial settings.

In this chapter, we propose the application of the YOLOv5[1] (You Only Look Once [179]) model to detect defects during industrial production, this paradigm aims to enhance the efficiency and accuracy of quality control, addressing the limitations of manual inspection and contributing to improved overall production quality. The subsequent sections will provide a detailed description of this paradigm, accompanied by experimental results to demonstrate the effectiveness of the proposed paradigm.

## 10.2    SUPERVISED LEARNING PARADIGM FOR EDIBLE OIL ANOMALY DETECTION

In this section, we will illustrate the paradigm with a detailed demonstration using the edible oil production industry as a case study. The defects within the interior of edible oil containers are black spot and hair, and we additionally detect the edible oil container to ensure accurate identification of the defects within the edible oil container interior. Figure 10.2 displays several examples of common edible oil defects.

---

[1]https://github.com/ultralytics/yolov5

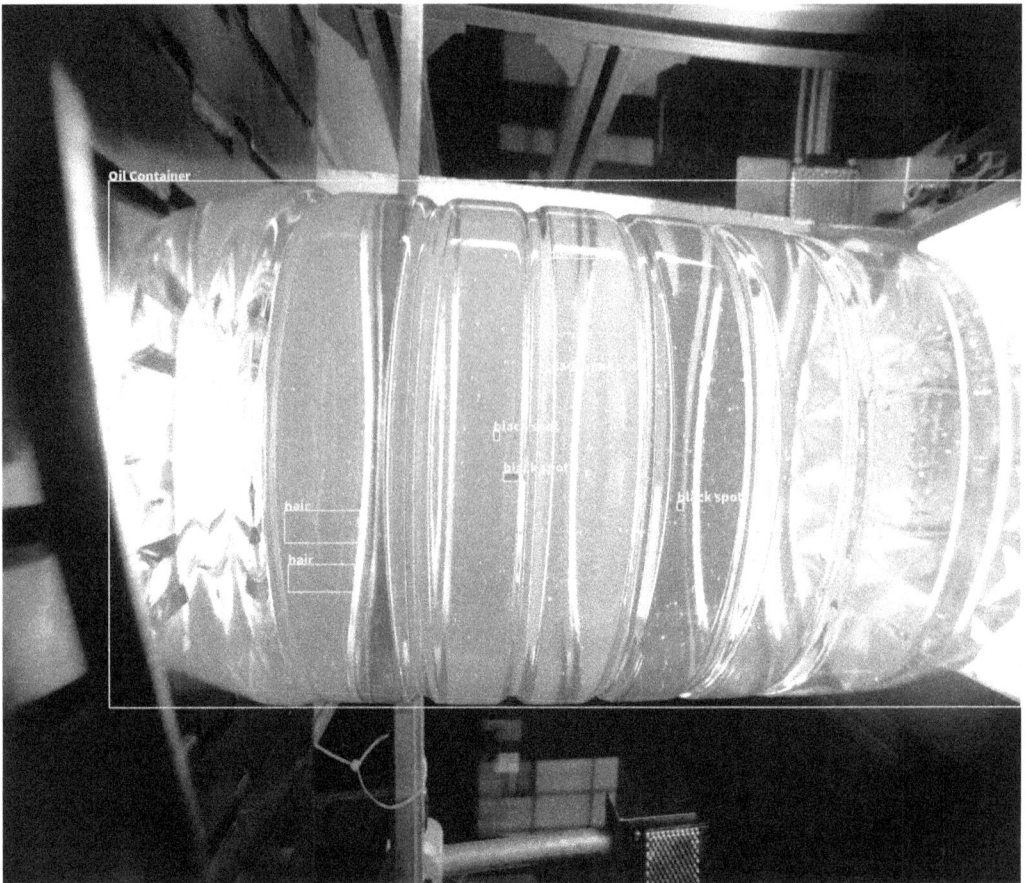

Figure 10.2   The image depicts typical targets for inspection: black spot and hair. Conducting inspections manually and visually, as per traditional methods, is not only time-consuming and labor-intensive but also highly susceptible to human error.

## 10.2.1   Data Acquisition

To effectively train and evaluate the YOLOv5 model, we construct a comprehensive dataset comprising high-resolution images with various sizes of two defects: black spot, and hair, and we additionally detect the edible oil container to ensure accurate identification of the defects within the edible oil container interior. The dataset is split into training, validation, and test sets. Annotations are meticulously created to ensure accuracy and completeness.

## 10.2.2   Image Pre-Processing

The images are pre-processed to enhance defect visibility and ensure compatibility with the YOLOv5 model.

### 10.2.2.1 Data Annotation

The dataset, which serves as the cornerstone for training and evaluating our defect detection model, must be meticulously annotated with bounding boxes that accurately delineate the regions of interest, namely, the defects present in the images. These annotations are the ground truth, and the benchmark against which our model's predictions will be measured.

To ensure the highest quality of annotations, each defect is carefully identified and outlined by experts. The bounding boxes are drawn in a way that completely encapsulates the defect without including unnecessary parts, thus providing a clear and unambiguous reference for the model. This precision is critical, as it directly impacts the model's ability to learn the distinguishing features of various defects. Moreover, the annotations are diverse, capturing a wide range of defect sizes to ensure the model's comprehensive understanding. This diversity is essential for the model to generalize well across different conditions, ultimately leading to a robust detection system.

The process of annotation also incorporates quality control measures to maintain the integrity and consistency of the dataset. This involves periodic reviews of the annotations and cross-verification by multiple annotators. By providing the dataset with such precise annotations, we establish a solid foundation for the model's training and evaluation. This, in turn, will enable us to develop a highly accurate and reliable edible oil anomaly detection system that can effectively identify and address the quality issues in the industry.

### 10.2.2.2 Data Augmentation

To enhance the robustness of the YOLOv5 model and improve its generalization capabilities, data augmentation techniques are applied to the training images. These techniques include rotation (randomly rotating the images to simulate different orientations of defects), flipping (horizontally and vertically flipping the images to increase variability), scaling (applying random scaling to create variations in defect sizes), contrast adjustment (modifying the contrast of the images to simulate different lighting conditions), etc. Moreover, considering that defects may appear at different scales, multi-scale processing of images [183] is performed to ensure that the model can detect defects of varying sizes.

Through these pre-processing steps, we can provide the model with a more standardized, diverse, and high-quality training dataset, thereby improving the accuracy and robustness of edible oil anomaly detection.

### 10.2.3 Defect Detection Using YOLOv5

YOLOv5 is an object detection model known for its speed and accuracy. It predicts bounding boxes and class labels for objects in an image in a single forward pass. The architecture of the YOLOv5 model used for edible oil anomaly detection is illustrated in Figure 10.3, which mainly includes Backbone, PANet [138] (Path Aggregation Network), and Output. Backbone is used to perform feature engineering from input

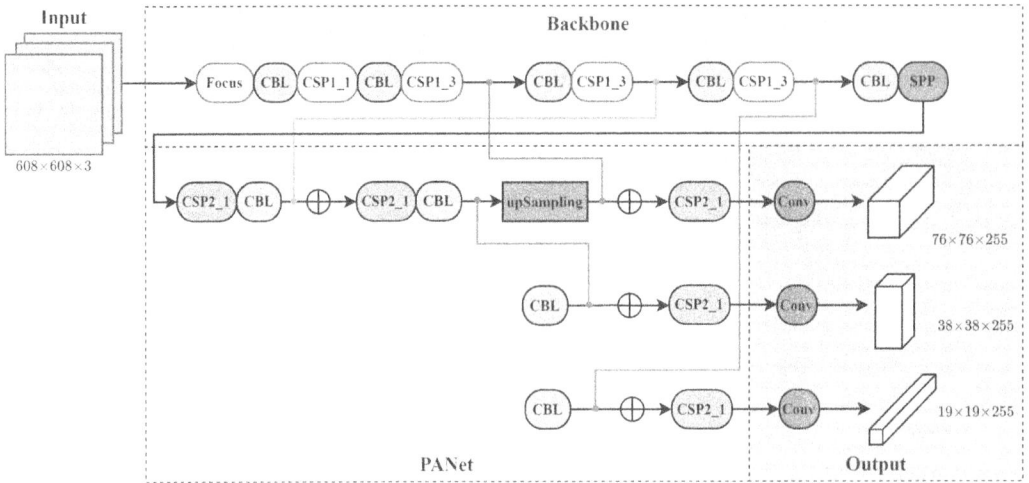

Figure 10.3   YOLOv5 architecture. ⊕ denotes concate operation.

images. PANet could obtain visual features robust to scale changes due to the used pyramid structure. The Output includes positions and labels of the regions of interest. The Positions, in the form of bounding boxes, are typically represented as rectangles, indicating the location of the target in the image, including the coordinates of the top-left and bottom-right corners.

### 10.2.3.1  Backbone

The backbone of YOLOv5 is responsible for extracting essential features from the input images. It typically uses a convolutional neural network [130] (CNN) to process the image through multiple layers, capturing different levels of abstraction. The backbone in YOLOv5 is based on the CSPDarknet53 [225] architecture, which includes:

- Convolutional Layers: These layers apply convolutional filters to the input image, generating feature maps that capture various aspects of the image.

- Batch Normalization: This technique normalizes the output of each convolutional layer to improve training stability and convergence.

- Residual Blocks: These blocks include skip connections that help in gradient propagation, enabling the network to learn deeper representations without the vanishing gradient problem.

### 10.2.3.2  PANet

The PANet is designed to enhance the feature maps produced by the backbone and create a rich representation of the input image. In YOLOv5, the PANet is designed to improve the flow of information throughout the network, particularly facilitating the aggregation of features from different layers, which is vital for detecting objects of various scales and complexities. Key features of PANet in YOLOv5:

- Bottom-Up Path Augmentation: PANet introduces a bottom-up path augmentation mechanism that complements the top-down pathway used in the Feature Pyramid Network (FPN). While FPN combines high-level semantic features with lower-level details, PANet's bottom-up path enhances feature maps by incorporating information from lower layers back into higher layers. This bi-directional information flow ensures that the network captures both detailed and abstract features effectively.

- Enhanced Information Propagation: By enabling feature maps from different network stages to interact, PANet improves the propagation of information across the network. This enhanced interaction allows for better contextual understanding and more accurate object localization.

- Improved Localization and Recognition: The combination of bottom-up and top-down pathways in PANet helps in better localization of objects and improves the recognition performance. This is particularly important for tasks where objects vary greatly in size and shape.

### 10.2.3.3 Output

The output of YOLOv5 is responsible for generating the final predictions, including bounding boxes, objectiveness scores, and class probabilities. The output processes the feature maps from the PANet and applies anchor boxes to predict the location and category of objects.

By leveraging the advanced features and robust architecture of YOLOv5, our paradigm aims to deliver superior performance in detecting common defects such as black spot and hair, thereby enhancing the quality control process in the industry.

## 10.2.4 Evaluation

The performance of the YOLOv5 model is evaluated using metrics such as precision, recall [17], F1-score, and mean Average Precision [79] (mAP). These metrics provide insights into the model's accuracy in detecting defects and its ability to minimize false positives and false negatives.

### 10.2.4.1 Precision and Recall

The precision-recall curve is used to analyze the trade-off between precision and recall at different detection thresholds, which reflects the algorithm's ability to distinguish between normal and defective images. Specifically, in our case, images with defects are treated as positive samples. Treating images with defects as positive samples is aimed at enabling the model to more accurately identify defective products, thereby enhancing the efficiency and precision of the quality inspection process. Precision (P) is calculated as follows:

$$precision = \frac{TP}{TP + FP} * 100\% \tag{10.1}$$

and recall (R) is calculated as follows:

$$recall = \frac{TP}{TP + FN} * 100\% \tag{10.2}$$

where TP (true positive) represents the number of samples whose labels are positive, and the actual forecasts are positive. FP (false positive) indicates the number of samples whose labels are negative, and the actual forecasts are positive. FN (false negative) represents the number of samples whose labels are positive, and the actual forecasts are negative. Based on the calculated P and R, the P-R curve could be obtained.

### 10.2.4.2 F1-score

F1 score is defined as the harmonic mean of precision and recall and calculated as follows:

$$F1 = \frac{2 * precision * recall}{precision + recall} \tag{10.3}$$

F1-score is a performance metric that balances precision and recall, providing a value to evaluate the overall effectiveness of a model. It is particularly useful when dealing with imbalanced datasets or when the costs of false positives and false negatives are different.

### 10.2.4.3 Mean Average Precision (mAP)

The mAP metric is computed to evaluate the overall performance of the model across all defect categories, which reflects the algorithm's ability to recognize specific defects. mAP represents the mean of different APs, where AP represents the area under the P-R curve. A high mAP value indicates that the model accurately detects and localizes defects in the samples. mAP is calculated as follows:

$$mAP = \frac{1}{K} \sum_{i=1}^{K} (R_i - R_{i-1}) P_i \tag{10.4}$$

where $K$ represents the number of categories.

## 10.3 EXPERIMENTS

### 10.3.1 Dataset

To effectively train and evaluate the YOLOv5 model for edible oil anomaly detection, we construct a comprehensive dataset comprising high-resolution images of samples with defects: black spot and hair, and we additionally detect the edible oil container to ensure accurate identification of the defects within the edible oil container interior. The dataset is split into training, validation, and test sets. Annotations are meticulously created to ensure accuracy and completeness.

The dataset comprises a total of 12,000 images, which are divided into three subsets: training, validation, and test sets. The division is done in a 70:15:15 ratio to ensure that the model has sufficient data for learning while also being able to validate and test its performance accurately:

- Training Set: Contains 8,400 images (70% of the total dataset).

- Validation Set: Contains 1,800 images (15% of the total dataset).

- Test Set: Contains 1,800 images (15% of the total dataset).

### 10.3.2   Training the YOLOv5 Model

The YOLOv5 model is trained on the training set using the Adam optimizer. The training process involves iterating through multiple epochs, during which the model parameters are adjusted to minimize the loss function. The learning rate, batch size, and other hyper-parameters are tuned to achieve optimal performance. The training hyper-parameters were as follows: epochs set to 50, batch size set to 20, learning rate set to 0.01, weight decay set to 0.0005, image size set to 1280, warmup_epochs set to 3.0.

### 10.3.3   Results

The trained YOLOv5 model is evaluated on the test set to assess its defect detection capabilities. Our model has demonstrated exceptional accuracy on the test set, achieving a mAP@0.5 of 0.96 and mAP@0.95 of 0.82. Notably, the model's ability to detect minute and subtle defects, such as black spots and hair, underscores the robustness and sophistication of our methodology. This indicates that the model has learned to accurately recognize patterns within the training data. This performance underscores the model's capability to meet real-world production requirements, validating the effectiveness of our approach. Example detection results are shown in Figure 10.4.

The impressive performance of the YOLOv5 model demonstrates its efficacy in edible oil anomaly detection. This high accuracy translates to tangible benefits for the industry. Firstly, the reduced risk of false positives ensures that only genuinely defective products are flagged, minimizing waste and rework while enhancing overall product quality. Secondly, the model's real-time processing capabilities enable instant detection and immediate corrective actions. Additionally, the consistent performance of YOLOv5 eliminates the variability associated with human inspectors, leading to more reliable quality control and a reduced risk of defective products reaching the market. The scalability and flexibility of YOLOv5 allow for its implementation across multiple production lines and adaptation to diverse defects and environments, offering a comprehensive and adaptable solution for quality control.

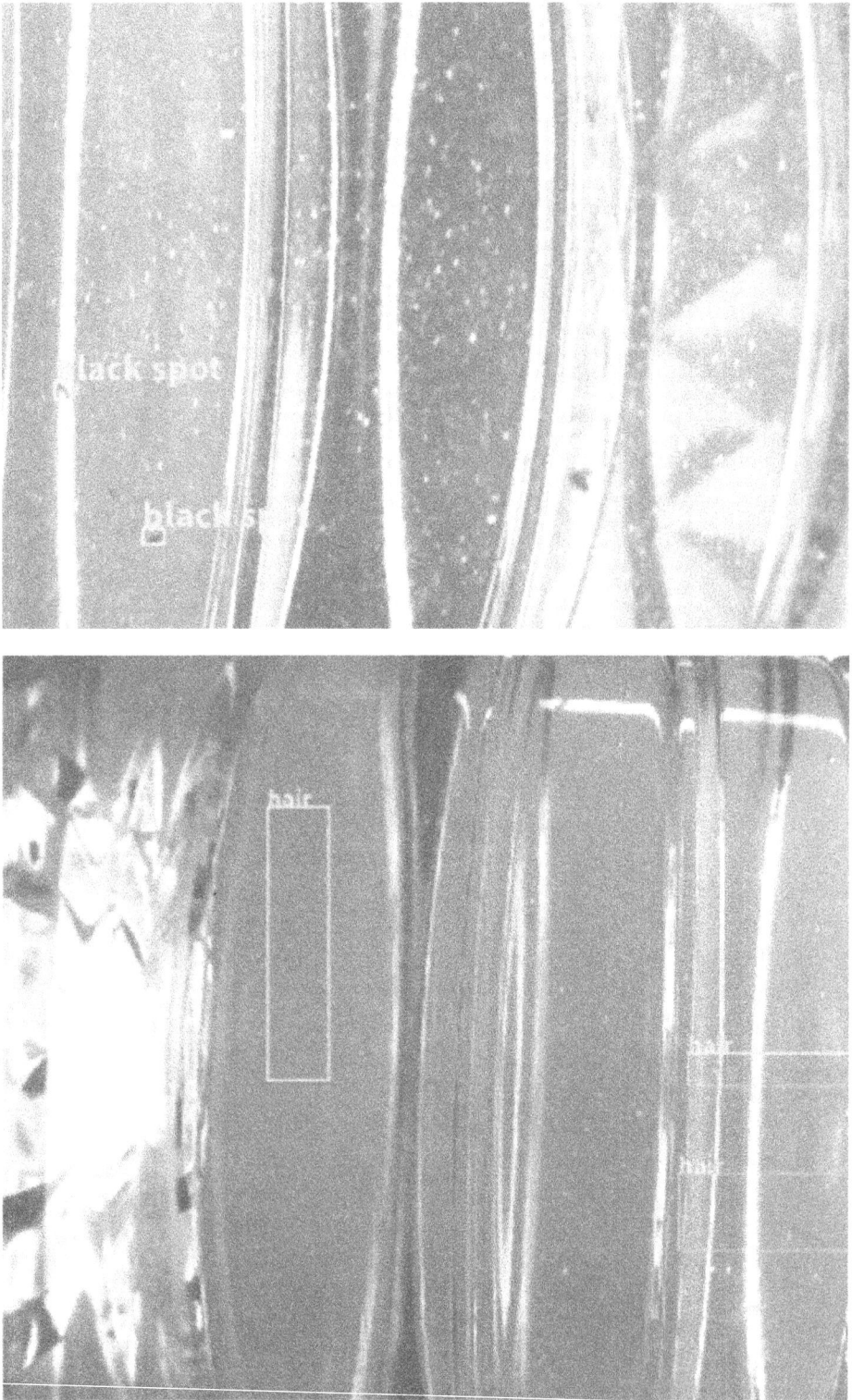

Figure 10.4 Example detection results from the YOLOv5 model. Defects are highlighted with bounding boxes.

## 10.4  CONCLUSION

This chapter proposes a supervised learning paradigm for industrial product quality inspection, which involves four main steps: data acquisition, image pre-processing, defect detection, and evaluation. By capturing images of the production line with cameras and employing a visual detection model for analysis, we automate and intellectualize the inspection process, substantially reducing the need for human intervention and enhancing overall efficiency and accuracy. The YOLOv5 model's application in edible oil anomaly detection represents a transformative leap in industrial quality control. Its automation of the defect detection process eliminates the need for labor-intensive manual inspections, offering significant time and cost savings. The model's high accuracy and consistency ensure reliable quality control, reducing the risk of products with defects reaching consumers. The scalability of the YOLOv5 model allows for its integration across multiple production lines, providing comprehensive defect monitoring and enabling timely corrective actions. The case study of edible oil has demonstrated the effectiveness of the proposed paradigm, which can be broadly applied to various fields, including electronics product quality inspection, metal material quality inspection, bearing quality inspection, glass cover plate quality inspection, etc. By leveraging this paradigm, manufacturers can achieve significant improvements in production efficiency, product quality, and cost-effectiveness.

# CHAPTER 11

# Unsupervised Learning Paradigm for Industrial Visual Anomaly Detection

## 11.1 INTRODUCTION

Quality inspection is critical in modern manufacturing industries because defective industrial products are unacceptable to downstream manufacturers and users and may even bring safety risks. Traditional manual inspection methods are time-consuming and susceptible to human error. With significant advancements in deep learning and computer vision [78], automated industrial visual anomaly detection has gained increasing attention across various industries. Typical approaches utilize supervised detectors (such as YOLOv5 [100]) or segmentation models (such as U-Net [187]), to locate and classify defects in captured images. While effective in some scenarios, these supervised methods have several limitations:

**Scarcity of Defect Samples.** In real-world scenarios, defect samples are often rare compared to normal samples, making it challenging for supervised models to learn effectively. Specifically, a limited number of defect samples leads to inadequate learning of their features, resulting in a decline in detection performance.

**High Cost of Annotated Data.** Supervised detectors or segmentation models require large datasets with accurately annotated defects, which are time-consuming, laborious, and expensive to obtain.

**Limited Generalization.** Supervised models typically learn only the features of defects present in the training set, which often leads to poor performance when encountering new or unseen defect types.

To address these limitations, this chapter proposes an unsupervised learning paradigm for industrial anomaly detection, which eliminates the need for defect annotations and enhances generalization to unseen anomalies. The chapter proceeds by detailing the unsupervised learning paradigm, followed by its application to battery and bearing anomaly detection, demonstrating its effectiveness through comprehensive experiments.

DOI: 10.1201/9781003644972-11

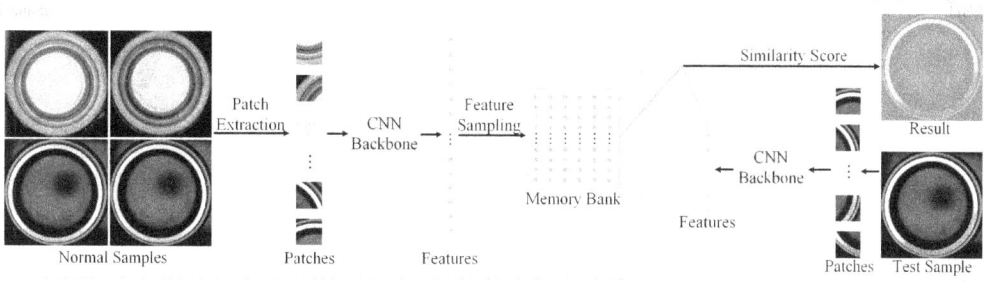

**Figure 11.1** The framework of the unsupervised paradigm for industrial visual anomaly detection.

## 11.2 UNSUPERVISED LEARNING PARADIGM

The unsupervised learning paradigm for industrial visual anomaly detection learns normal patterns and features only from samples without defects effectively and considers patterns that are not similar to normal patterns as anomalies. The overall framework of the paradigm is illustrated in Figure 11.1, it includes both training and testing processes.

### 11.2.1 Training

Before testing, the unsupervised anomaly detection network should be trained using only normal samples. The goal of the training phase is to construct a memory bank $M$ that stores robust and representative features of normal samples. The training process involves the following steps:

#### 11.2.1.1 Normal Sample Collection

High-resolution images of products without defects are collected to form a dataset of normal samples, denoted as:

$$S_N = \{(x_1, y_1), (x_2, y_2), \ldots, (x_n, y_n)\} \tag{11.1}$$

Where $x_i$ is the image of $i_{th}$ normal sample, and $y_i \in \{0, 1\}$ indicates its label. $y_i = 0$ indicates there is no defect in $x_i$, and $y_i = 1$ indicates the presence of defects in $x_i$. So, for the normal sample set $S_N$, $\forall i \in [1, n], y_i = 0$.

#### 11.2.1.2 Patch Extraction

Each normal image $x_i$ is divided into $k$ smaller, overlapping patches $p_i = (p_i^1, p_i^2, \ldots, p_i^k)$. The size and overlap of patches are chosen to balance detailed local analysis and computational efficiency. Patches of all images in $S_N$ construct a set:

$$P = \left\{ \left( p_1^1, p_1^2, \ldots, p_1^k \right), \left( p_2^1, p_2^2, \ldots, p_2^k \right), \ldots, \left( p_n^1, p_n^2, \ldots, p_n^k \right) \right\} \tag{11.2}$$

### 11.2.1.3 Feature Extraction

For all image patches of normal samples, we extract rich features and produce a set of feature vectors:

$$F = \left\{ \left( f_1^1, f_1^2, \ldots, f_1^k \right), \left( f_2^1, f_2^2, \ldots, f_2^k \right), \ldots, \left( f_n^1, f_n^2, \ldots, f_n^k \right) \right\} \tag{11.3}$$

To avoid features too generic (low representation) or too abstract (low sensitivity to small defects), we recommend using mid-level features of convolutional neural networks.

### 11.2.1.4 Memory Bank Construction

Since many normal image patches have similar visual features, the feature set $F$ contains a substantial amount of redundant features. To enhance the efficiency of our unsupervised paradigm, it is essential to select a subset of representative feature vectors from $F$ to construct a memory bank $M$, which will be used for similarity comparison during the testing phase.

## 11.2.2 Testing

In the testing phase, we detect and locate defects in test images using the trained memory bank $M$. The testing process involves the following steps:

### 11.2.2.1 Patch Feature Extraction

Similar to training, the input test image is divided into overlapping patches, and feature vectors are extracted from each patch. The method of patch cropping and feature extraction remains consistent with that of the training process.

### 11.2.2.2 Similarity Calculation

The similarity scores between the feature vector of each test image patch and all the feature vectors in $M$ are calculated.

### 11.2.2.3 Defect Localization and Scoring

Patches with lower similarity scores than specific score thresholds are considered to contain defects. For each test image, the recognition results of all patches are aggregated to output the anomaly result, highlighting defects in the test image.

## 11.3 IMPLEMENTATION AND EXPERIMENTS

To further introduce and validate our proposed paradigm, we implement it for the anomaly detection of battery and bearing. First, we construct a dataset comprising a training set and a test set. Next, we describe the implementation details and perform the anomaly detection task according to the paradigm.

(a) Normal samples

(b) Samples with defects

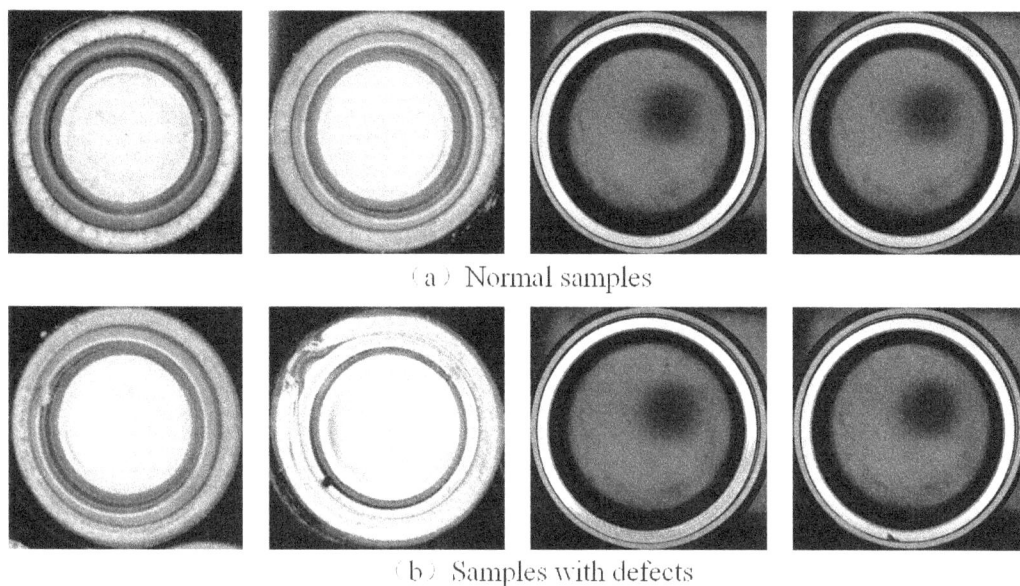

Figure 11.2  Example images from the dataset. (a) battery and bearing images of normal samples. (b) Images of samples with defects. Columns 1 and 2 show batteries, and columns 3 and 4 show bearings.

## 11.3.1  Dataset

Inspired by the public dataset, MVTec [11], we collect images from real-world battery and bearing samples and construct a dataset to evaluate our method. The dataset includes 2500 images of battery and 2500 images of bearing. A subset of these images contains annotated defects (500 battery images and 500 bearing images with various types of defects). All images are captured at a resolution of 640x640, and corresponding defects are annotated at pixel level. The dataset is split into a training set containing 70% normal images (1400 battery images and 1400 bearing images) and a test set comprising the remaining images with or without defects. Example images from the dataset are shown in Figure 11.2.

## 11.3.2  Implementation Details

In the training phase, we first divide training images into 64x64 pixel patches with 50% overlap. We then employ ResNet50 [78], pre-trained on the ImageNet [42] dataset, as the feature extractor. Specifically, the features of each patch are aggregated by the features of Layer 2 and Layer 3 of ResNet50, this is consistent with PatchCore [190]. Finally, The core-set sampling method in PatchCore [190] is utilized to select a diverse and representative subset of $F$ to construct the memory bank $M$.

In the testing phase, we begin by extracting the mid-level feature vectors of the test image patches following the approach used in the training phase. Next, we calculate the similarity scores between feature vectors of the test image patches and

TABLE 11.1   AUROC (image level, pixel level) for our unsupervised anomaly detection approach on the test dataset.

| Product Type | AUROC (Image Level) | AUROC (Pixel Level) |
|---|---|---|
| Battery | 96.1 | 85.5 |
| Bearing | 90.4 | 80.7 |

the vectors in $M$ using cosine similarity. Finally, we apply similarity thresholds to classify regions of the test image as either normal or anomalous.

### 11.3.3   Experimental Results

To analyze the performance of our paradigm, we quantitatively compute the Area Under the Receiver Operating Characteristic Curve (AUROC) [244] at both image and pixel levels, and visualize the qualitative results.

#### 11.3.3.1   Quantitative Results

Similar to [10, 35, 190], we evaluate the performance of our paradigm using AUROC at image level and pixel level. Higher AUROC scores indicate higher detection accuracy. The results are summarized in Table 11.1.

 The results demonstrate the effectiveness of our paradigm in detecting defects in real-world industrial anomaly detection scenarios.

#### 11.3.3.2   Qualitative Results

In addition to the quantitative metrics, we provide visualizations to demonstrate the effectiveness of our method in Figure 11.3. The first row shows input images, the second row shows ground truth labels, and the third row shows the detected anomalies. Our method effectively identifies and localizes defects in both battery and bearing samples.

 By combining quantitative and qualitative assessments, we demonstrate the robustness and effectiveness of the unsupervised learning paradigm for industrial visual anomaly detection. Importantly, when applying our paradigm to other industrial visual anomaly detection scenarios, it is only necessary to collect an appropriate amount of normal sample images and train anomaly detectors in the same way. This general unsupervised learning paradigm significantly advances industrial visual anomaly detection by overcoming the drawbacks of supervised methods.

## 11.4   CONCLUSION

This chapter introduces a general unsupervised learning paradigm for industrial visual anomaly detection, designed to address the challenges of defect scarcity, high

（a）**Battery**

（b）**Bearing**

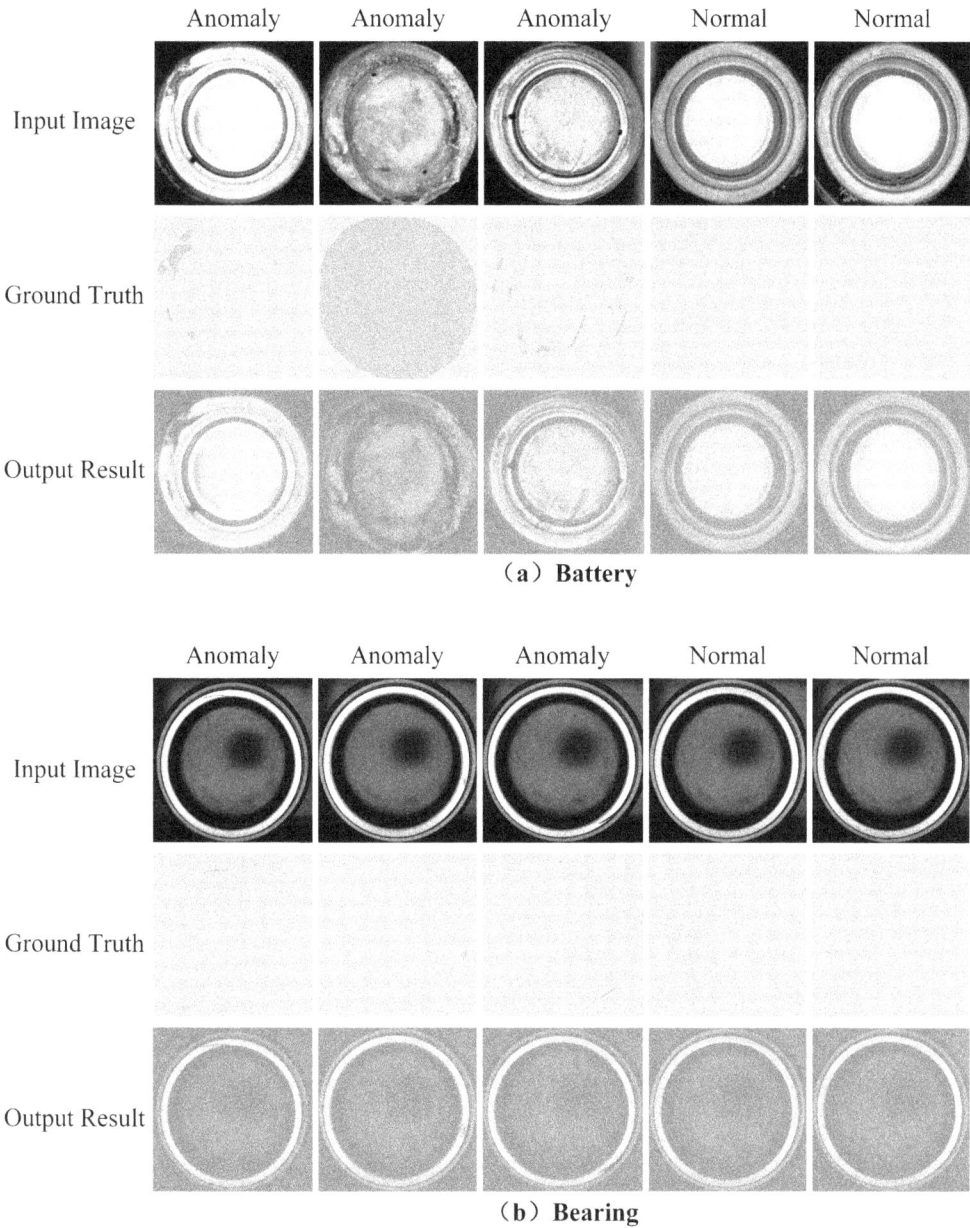

Figure 11.3   Visualization of detection results on the test dataset. (a) Results on battery data. (b) Results on bearing data. For each subgraph, the first three columns represent defective samples, while the last two columns represent normal samples in the test dataset.

annotation costs, and limited generalization in supervised methods. The paradigm extracts features from only normal samples, stores them in a memory bank, and detects potential defects by comparing the similarity of test image patches to those in the memory bank. Taking two representative industrial anomaly detection scenarios, battery anomaly detection and bearing anomaly detection, as examples, we implement the proposed unsupervised paradigm and demonstrate its effectiveness through comprehensive experiments. Beyond these applications, the paradigm is adaptable to other industrial visual anomaly detection scenarios, such as fabric anomaly detection, steel anomaly detection, and engine anomaly detection, etc.

# CHAPTER 12

# Object Identity Recognition Paradigm for Fishing Boat Recognition

## 12.1  INTRODUCTION

The demand for Object Identity Recognition (OIR) spans multiple industries. For example, in the insurance sector, verifying livestock identity—such as that of cattle and sheep—is essential to prevent insurance fraud. Similarly, in marine fisheries, identifying fishing boats is critical for combating fraudulent vessel registrations. At present, these OIR tasks primarily depend on manual processes, which are both time-consuming and costly.

To overcome these challenges, we introduce a novel OIR paradigm. This paradigm utilizes advanced computer vision techniques to streamline and automate the identity recognition process. Drawing inspiration from face recognition methodologies, our approach involves the following steps:

**Object Detection:** Identifying the object of interest in an image or video.

**Key Points Detection:** Locating specific key points on the object to facilitate further analysis.

**Alignment:** Aligning the object based on the detected key points to ensure consistency in feature extraction.

**Feature Extraction:** Extracting features from the aligned object image.

**Similarity Comparison:** Comparing the extracted features with the stored features in the database to verify the identity of the object.

The proposed paradigm seeks to replace manual processes, significantly improving efficiency while reducing costs. Automation mitigates the limitations of manual methods, enabling more effective management and enforcement across various domains.

In this chapter, we focus on a specific application of the proposed paradigm: fishing boat recognition.

DOI: 10.1201/9781003644972-12

## 12.2 FISHING BOAT RECOGNITION

At present, fishing boat recognition relies solely on manual processes. Specifically, after a fishing boat docks, officers board the vessel to verify whether its engine number matches the registered number. If the numbers match, the boat is deemed legally registered; otherwise, it is flagged as using a fraudulent identity. This manual verification process is time-consuming, requiring approximately 40 minutes per boat, and prone to errors due to factors such as complex port environments, the high structural and visual similarity of fishing boats, and disorganized registration by fishermen. The shortage of personnel, outdated evidence collection methods, and lengthy, inefficient workflows have contributed to widespread violations, including illegal offshore fishing, fraudulent boat identities, and unlicensed fishing operations. These issues undermine the sustainable management of marine resources, pose serious safety risks to fishing operations, and disrupt the effective functioning of fisheries management authorities.

Our proposed OIR paradigm for fishing boat recognition, as depicted in Figure 12.1, begins with optical character recognition (OCR) to extract the boat number (or manual input if necessary) and retrieve the corresponding cabin image from the database. Next, the fishing boat recognition process begins by capturing an image of the boat, detecting the cabin area, and identifying its key points. The cabin area is then classified, and a key point template is applied to align the region. Finally, features are extracted from the aligned cabin image and compared against the database entry associated with the boat number. If the similarity score surpasses the predefined

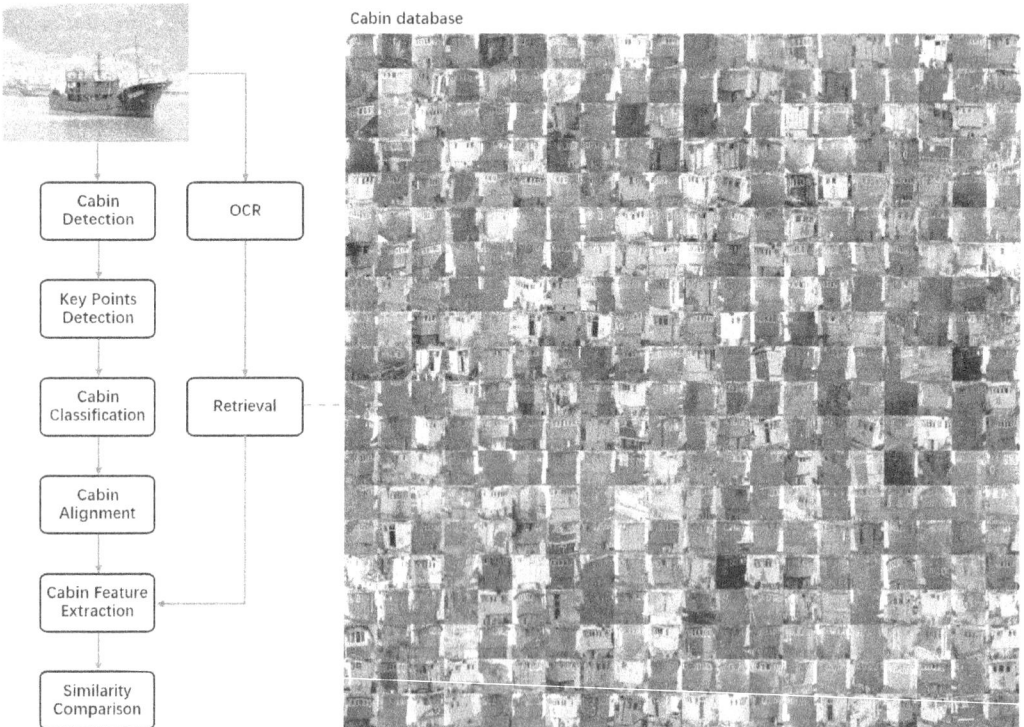

Figure 12.1   The process of fishing boat recognition.

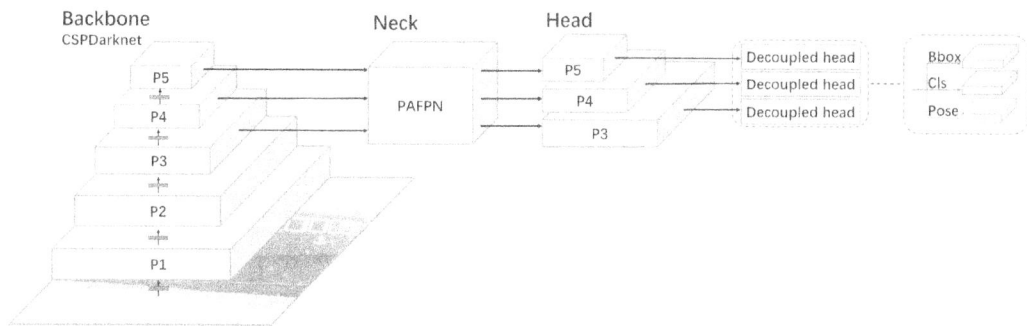

Figure 12.2  YOLOv8 model structure. In the decoupled-head module, adding a key points detection branch (Pose branch) enables the model to simultaneously perform both object detection and key points detection tasks.

threshold, the boat's identity is verified successfully; otherwise, it is identified as having a fraudulent identity.

## 12.2.1  Cabin Detection

To minimize the influence of background information on the verification of fishing boats, we define the cabin as the critical part of the boat. We used the object detection method YOLOv8 (an improved version of the You Only Look Once [179] method) to detect the cabin on the fishing boat.

As shown in Figure 12.2, the YOLOv8 model consists of three main components: the backbone, neck, and head. The backbone is built upon an enhanced CSPDarknet53 [181] architecture, which improves feature extraction and efficiency. The neck incorporates Path Aggregation Feature Pyramid Network (PAFPN) [138] modules to achieve multi-scale feature fusion. The head adopts a decoupled-head [62] structure, separating classification and detection tasks, and transitions from an anchor-based to an anchor-free [264] approach. Overall, YOLOv8 is designed for efficiency, ensuring high-precision detection.

Figure 12.3 presents the results of cabin detection, where the dark box highlights the detected cabin area.

## 12.2.2  Key Points Detection

To mitigate the interference caused by different viewing angles during the identification of fishing vessels, it is essential to align the images using key points prior to the recognition process. We used YOLOv8 for key points detection. As shown in Figure 12.2, the addition of a key points detection branch (Pose branch) in the decoupled-head module enables the model to detect objects and key points simultaneously.

In our scenario, a cabin includes three faces: front, left, and right, with four corner points extracted as key points for each face. Therefore, a cabin has a total of 12 key points. When key points for certain faces are not visible, we represent them

Figure 12.3 The result of cabin detection and key points detection. The cabin and its key points are clearly detected, with dark box highlighting the cabin's detection and four grey dots marking left-side key points, while two white triangles indicate visible front key points.

as (-1, -1). Additionally, the presence of all four key points on a specific face allows us to determine whether the image shows the front, left, or right face of the cabin.

As depicted in Figure 12.3, the four key points located on the left side of the cabin are marked with grey dots. In contrast, the two visible key points on the front face are highlighted with white triangles.

### 12.2.3 Cabin Classification

Significant variations exist among different cabins, making it impractical to apply a single template with fixed key points for image alignment. To address this challenge, cabins were categorized into 24 distinct classes based on their exterior color, shape, and other features. For each category, tailored key point templates were applied to the front, left, and right faces of the cabins. Figure 12.4 illustrates representative examples of various categories for the left faces of cabins.

To classify the cabins, two ResNet50-based classification models were trained. The first model identifies the left and right faces of the cabins, while the second is dedicated to recognizing the front face.

### 12.2.4 Cabin Alignment

By analyzing the key points to identify whether the cabin image showed the front, left, or right face, and using the classification model to determine the cabin category,

Figure 12.4  The categories for the left faces of the cabins.

we selected the appropriate key point template. The image alignment process utilized a four-point perspective transformation, as detailed in Equation (12.1).

$$\begin{bmatrix} x_i' \\ y_i' \\ 1 \end{bmatrix} = H \cdot \begin{bmatrix} x_i \\ y_i \\ 1 \end{bmatrix} \quad (12.1)$$

where $H$ is:

$$H = \begin{bmatrix} h_{11} & h_{12} & h_{13} \\ h_{21} & h_{22} & h_{23} \\ h_{31} & h_{32} & h_{33} \end{bmatrix}$$

Figure 12.5  The result of cabin alignment.

And $(x_i, y_i)$ represents the pixel coordinates in the original image. $(x'_i, y'_i)$ represents the pixel coordinates in the template image.

By utilizing four corresponding points from the original image and the template image, the transformation matrix $H$ can be computed.

As illustrated in Figure 12.5, the cabin area has been successfully aligned, demonstrating the effectiveness of the perspective transformation.

### 12.2.5  Cabin Feature Extraction

ResNet50 is used to extract 512-dimensional features from the aligned cabin images. The model is trained using the ArcFace [43] method, a technique widely adopted in facial recognition.

### 12.2.6  Similarity Comparison

The cosine similarity between features extracted from two aligned cabin images is calculated. If the similarity value exceeds a threshold, empirically determined to be 0.45 in this study, the two images are classified as belonging to the same boat; otherwise, they are classified as different. As illustrated in Figure 12.6, the similarity score between the two aligned cabins is 0.51, suggesting that the cabins are likely from the same boat.

## 12.3  EXPERIMENTS

### 12.3.1  Dataset

Our dataset consists of cabin collections obtained through three methods: smartphone photography, smartphone video recording, and drone video recording. The initial dataset includes 2,755 fishing boats. For each boat, we collected at least 50 images capturing various cabin positions and 50 images of the entire boat, all taken from diverse angles. This resulted in a total of approximately 300,000 images.

Similarity: 0.51

Figure 12.6   The similarity of two aligned cabins.

## 12.3.2   Data Annotation

**Object Detection:** Each image was manually annotated with a bounding box indicating the cabin area.

**Key Points Detection:** The visible key points for the front, left, and right faces of the cabin were manually annotated.

**Cabin Classification:** After extracting the cabin area, we categorized it into 24 distinct classes based on features such as color and shape. These annotations were utilized to train the cabin classification model.

**Feature Extraction:** For each cabin area, the appropriate template was selected based on its category and corresponding face. The area was then aligned using a four-point projection transformation, and the aligned images were used for training the feature extraction model.

## 12.3.3   Training

The YOLOv8 model was trained using both object detection and key points annotation data. The hyperparameters were as follows: 100 epochs, a batch size of 16, a learning rate of 0.01, a weight decay of 0.0005, and an image size of 1280.

In the cabin classification task, Two ResNet50-based classification models were trained—one for the left and right faces of the cabins and another for the front face. The hyperparameters were as follows: 40 epochs, a batch size of 256, a learning rate of 0.1, a weight decay of 0.0005, and an image size of 224.

For the feature extraction task, the ResNet50 model was trained with the ArcFace method. The hyperparameters were: 20 epochs, an embedding size of 512, a batch

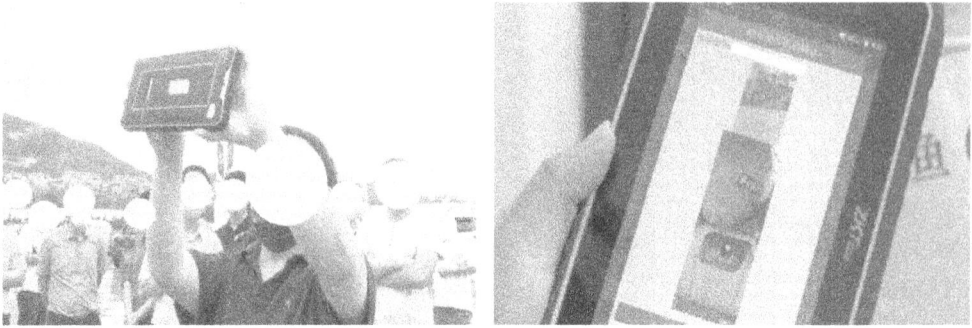

Figure 12.7   The practical application of the fishing boat recognition system.

size of 64, a learning rate of 0.01, a weight decay of 0.0005, margin parameters s: 64, m1: 1.0, m2: 0.5, m3: 0.0, and an image size of 224.

### 12.3.4   Results

The YOLOv8 model achieved a mean average precision (mAP) of 0.67 for cabin detection and 0.73 for cabin key points detection on the validation set. The cabin classification models achieved accuracies of 97.8% for the left and right faces and 96.3% for the front face.

Figure 12.7 illustrates the practical application of our fishing boat recognition system. To date, the system has registered 3,100 local fishing boats. Over the past month, statistical analysis of verification records shows that the system achieved an accuracy rate of 86% in boat verification. Moreover, the system significantly reduced the verification time per boat, allowing officers to complete the process within just a few minutes.

## 12.4   CONCLUSION

This chapter introduces a novel OIR paradigm that revolutionizes OIR processes across industries through advanced computer vision techniques. The paradigm encompasses a comprehensive framework that includes object detection, key points detection, alignment, feature extraction, and similarity comparison, automating the OIR process to reduce costs and improve efficiency significantly.

We demonstrated the practical application of this paradigm through a case study on fishing boat recognition. By leveraging YOLOv8 for cabin detection and key points identification, and ResNet50 for cabin classification and feature extraction, our system achieved an 86% accuracy rate and reduced the traditional boat verification time from 40 minutes to just a few minutes. This approach enhances fisheries management by addressing issues such as fraudulent boat identities and unlicensed fishing operations.

Beyond fishing boat recognition, the proposed paradigm shows strong potential for applications in other industries. For instance, it could be adapted for livestock identification in insurance or vehicle identification for regulatory compliance. This robust framework provides a scalable solution to improve efficiency and reduce costs in diverse OIR scenarios.

# CHAPTER 13

# Image Editing Paradigm for Clothing Fashion Customization

## 13.1  INTRODUCTION

Fashion design has always been a dynamic and creative field, where customization and personalization play pivotal roles in meeting individual preferences. As digital tools become more integrated into the design process, the demand for efficient and flexible methods for virtual clothing modification continues to grow. Existing techniques, such as DragGAN [163], enable image transformation by adjusting global features, such as garment shapes and overall styles. However, they lack the ability to fine-tune intricate details like colors, patterns, and textures, which are essential in fashion.

We propose a novel image editing paradigm that overcomes these limitations by combining shape modification capabilities with cutting-edge segmentation and inpainting models. By incorporating accurate segmentation of garment areas and high-quality inpainting, this paradigm allows for detailed customization of both the garment's shape and its finer attributes, such as fabric texture and color patterns.

The key idea behind this paradigm is its modular approach to clothing customization, which can be broken down into three distinct stages:

- **Interactive Shape Modification**: This initial stage allows users to manipulate the overall shape and structure of the garment. By interactively adjusting control points, users can modify elements like sleeve length, neckline style, or overall silhouette, providing a high degree of flexibility in achieving the desired design.

- **Semantic Segmentation**: Following the shape modification, the system employs a semantic segmentation technique to precisely identify and isolate the modified garment area. This ensures that subsequent modifications are accurately targeted and applied only to the intended region, maintaining the integrity of the original image.

DOI: 10.1201/9781003644972-13

- **Inpainting for Detailed Customization**: In the final stage, an inpainting model is used to add intricate details and refine the visual appearance of the garment. Users can specify their preferences for aspects like color, texture, pattern, and fabric type, allowing for a highly personalized and aesthetically pleasing result.

The innovation lies in the seamless integration of these technologies, enabling users to alter not just the overall form of a garment but also its aesthetic details, offering unprecedented creative freedom in virtual garment design. This chapter demonstrates the application of the paradigm through a specific implementation using DragGAN, SAM [249], and SDXL [125] to showcase its effectiveness in clothing customization.

## 13.2  METHODOLOGY

The proposed methodology consists of several stages that work together to allow for both shape manipulation and detailed customization of virtual clothing. These stages include shape modification, segmentation, and inpainting, with each phase contributing to the final garment design. The workflow is shown as 13.1:

Figure 13.1 illustrates the comprehensive workflow of our virtual clothing approach, which combines DragGAN, SAM, and SDXL inpainting to achieve precise and realistic results. The process begins with DragGAN, which is employed to manipulate and reshape specific elements of the clothing, such as adjusting the sleeves' length or position. SAM is then used to generate a precise mask of the modified clothing item, ensuring accurate localization of the targeted area. The segmented mask, along with the original image and a masked version, serves as input for the SDXL inpainting model. Guided by a textual description provided by the user, SDXL

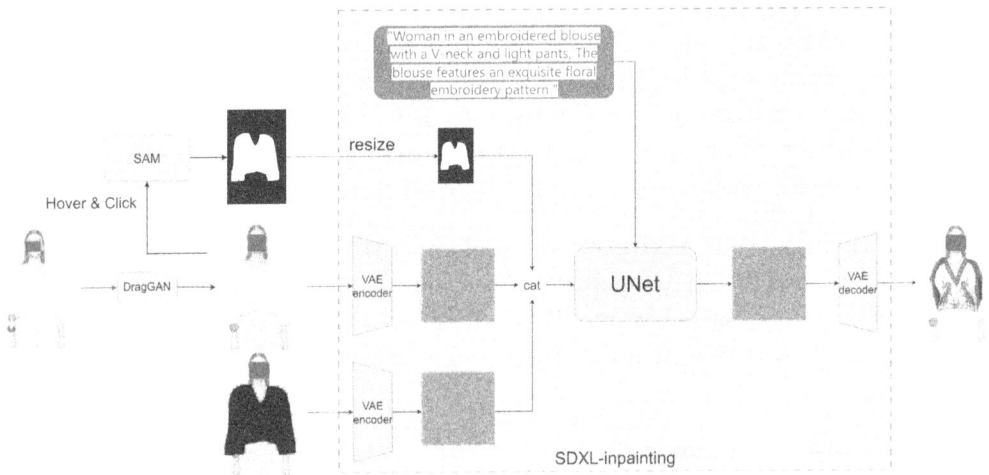

Figure 13.1  Pipeline of virtual clothing using DragGAN, SAM, and SDXL inpainting. The process integrates shape adjustment, mask generation, and detailed image synthesis guided by textual descriptions.

latent code w                    w'                                    w*

Generator    Motion        Generator    Motion              Generator
             supervision                supervision

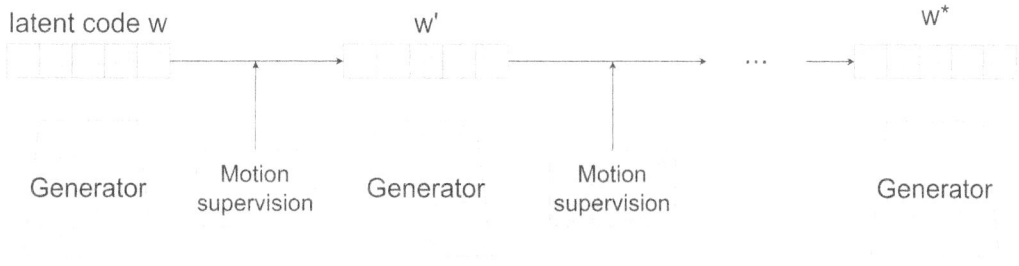

Figure 13.2   The pipeline of DragGAN. DragGAN enables interactive image manipulation by allowing users to specify handle and target points, which the system moves precisely through feature-based motion supervision. The motion supervision is achieved via a shifted patch loss on the feature maps of the generator.

synthesizes a detailed and visually coherent modified garment, integrating elements such as color, patterns, and fabric textures. This integrated pipeline effectively combines geometric manipulation, semantic segmentation, and generative modeling to produce customized clothing designs while preserving the realism of the overall image.

## 13.2.1   DragGAN: Interactive Shape Modification

DragGAN is a generative model that enables precise, interactive image manipulation. Unlike traditional methods that require complex tools or direct pixel-level editing, DragGAN allows users to modify an image's global structure by simply dragging control points. This makes it particularly useful for tasks like adjusting the overall shape or fit of clothing items, such as resizing sleeves, altering the cut of a dress, or reshaping the silhouette of pants.

DragGAN, shown in Figure 13.2, enables interactive image manipulation by allowing users to specify handle and target points, which the system moves precisely through feature-based motion supervision. It leverages the discriminative features of GANs to track points and optimize latent codes, ensuring realistic deformations on the generative image manifold. DragGAN operates by defining specific control points on the image. These points can be placed on key areas of the garment (e.g., collar, waistline, hem) and then dragged to new positions. The model then generates a new image with the shape adjusted according to these movements.

The underlying architecture is a Generative Adversarial Networks (GAN) [71], which consists of a generator that creates realistic images and a discriminator that ensures the output matches the original dataset's style and content. This setup guarantees that the shape modifications look natural while maintaining visual consistency with the rest of the image.

The key advantage of DragGAN is its real-time, interactive interface. Users can directly manipulate the image, seeing immediate feedback as the image adapts to the control points.While DragGAN excels at global shape transformations, it does not modify the finer details of the image, such as textures, patterns, or colors. For this reason, additional processes are required to refine the garment's appearance.

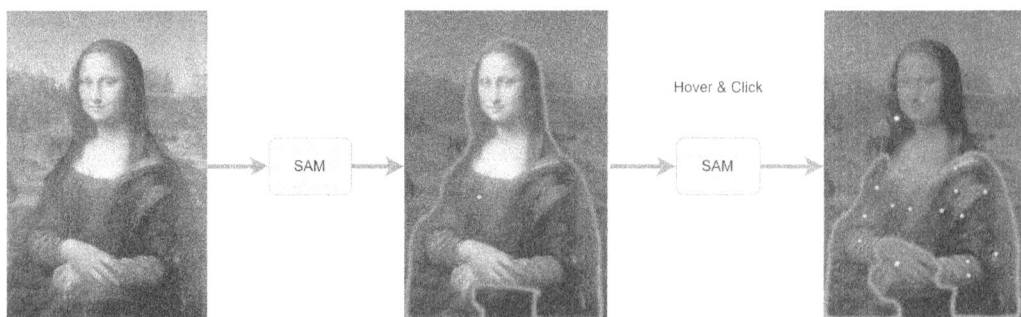

Figure 13.3  A showcase of SAM's operation. SAM's user interface offers intuitive tools for manual refinement and adjustment of the segmentation mask. Users can easily edit the mask to ensure it accurately captures the modified garment area.

DragGAN is limited in its ability to make finer adjustments, such as changing fabric textures or applying new color schemes, which is essential for virtual clothing design. These adjustments need to be handled by subsequent models in the pipeline.

## 13.2.2  SAM: Semantic Segmentation for Target Isolation

While DragGAN offers powerful shape manipulation capabilities, it lacks the ability to precisely isolate the modified garment area for targeted modifications. This is where the Segment Anything Model (SAM) [111] comes into play. SAM is a cutting-edge semantic segmentation tool that excels at identifying and delineating objects within an image, even in complex scenes with multiple overlapping elements. By leveraging SAM, we can achieve pixel-level accuracy in isolating the garment area, ensuring that subsequent modifications are applied only to the intended regions.

SAM's deep learning architecture is specifically designed to recognize and differentiate between various objects, including clothing items, in an image. This allows it to accurately identify and segment the modified garment, even in cases where it overlaps with other objects like the body or background. SAM's ability to handle diverse garment shapes, sizes, and orientations makes it a versatile tool for various clothing items, from flowing dresses to tailored suits.

SAM's user interface offers intuitive tools for manual refinement and adjustment of the segmentation mask, shown in Figure 13.3. Users can easily edit the mask to ensure it accurately captures the modified garment area, even in cases where the automatic segmentation is not perfect. This level of control is crucial for achieving precise modifications and maintaining the overall visual coherence of the image.

The segmented mask generated by SAM serves as a crucial input for the subsequent stages of the pipeline, guiding the inpainting process performed by SDXL. By ensuring that only the modified garment area is targeted for inpainting, SAM helps maintain the integrity of the original image while allowing for detailed customization of the garment itself.

SAM's robustness to various image conditions, including lighting, background complexity, and garment occlusions, ensures reliable segmentation even in challenging scenarios. The efficiency of SAM's segmentation process minimizes the overall time required for the virtual clothing modification workflow, making it suitable for real-time applications.

Consider a user who has used DragGAN to adjust the neckline of a dress in an image. SAM can then be employed to accurately segment the newly modified neckline area, even if it overlaps with the wearer's body or other parts of the dress. This precise segmentation allows SDXL to focus on inpainting the new neckline design with the desired fabric texture, color, and pattern, ensuring a seamless and realistic result. In summary, integrating SAM into the virtual clothing modification pipeline enhances the accuracy, control, and efficiency of the process, enabling users to achieve their desired modifications with ease and precision.

### 13.2.3  SDXL: High-Fidelity Inpainting for Detailed Customization

The final stage of our framework leverages the SDXL (Stable Diffusion XL) [169] model's powerful inpainting capabilities to generate the desired clothing designs based on user specifications. This process involves a sophisticated integration of latent-space manipulation, advanced generative modeling, and user-defined guidance. By combining the modified image from DragGAN, the mask generated by SAM, and additional user inputs, the system achieves highly detailed and personalized clothing transformations.

While DragGAN handles global shape changes and SAM isolates the garment, SDXL is responsible for adding intricate details like patterns, textures, colors, and fabric finishes based on user input. This is where the final garment design is fully realized. SDXL uses text prompts to guide the inpainting process, allowing users to specify detailed modifications.

SDXL is built upon the concept of diffusion models [38], which iteratively refine an image by reversing the diffusion process. SDXL , shown in Figure 13.4, employs a two-stage process where a base model generates an initial image based on a given prompt. The refiner model then enhances the image quality and details, ensuring a

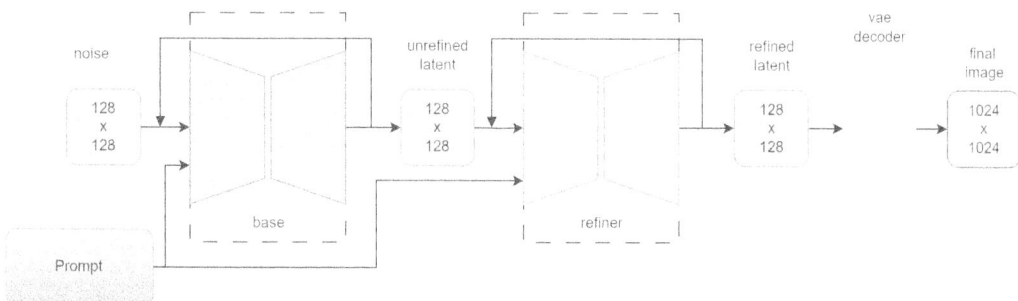

Figure 13.4  Visualization of the SDXL pipeline. SDXL employs a two-stage process where a base model generates an initial image based on a given prompt. The refiner model then enhances the image quality and details, ensuring a higher resolution and more refined output.

higher resolution and more refined output. Starting with random noise, the model progressively denoises the image, incorporating the specified text prompts at each step. This ensures that the inpainted details are consistent with both the garment's overall structure and the desired aesthetic.

SDXL is known for its ability to generate highly detailed and photorealistic images. It can create complex patterns, realistic textures (such as wool, cotton, silk), and intricate details (like stitching, buttons, or embroidery), making it ideal for fashion and garment design.

### 13.2.3.1   VAE: Latent Representations

At the core of this process is the manipulation of latent representations, which serve as the intermediary format between the input and output images in the SDXL pipeline.

The DragGAN-modified image is first encoded into a latent representation using the SDXL model's variational autoencoder (VAE) [168]. The VAE consists of an encoder that compresses high-dimensional input data (e.g., the modified image) into a lower-dimensional representation and a decoder that reconstructs the image from this compressed representation. This architecture enables efficient data processing and ensures that important features (e.g., shape and texture) are maintained. By reducing the dimensionality of the data, the VAE streamlines the pipeline, ensuring that the inpainting process can focus on generating details without being overwhelmed by excessive input complexity.

Specifically, this encoder compresses the high-dimensional DragGAN-modified image into a more compact and feature-rich representation with a channel dimension of 4. Similarly, a masked-out version of the DragGAN-modified image—where the masked region is replaced with a placeholder—is also processed through the VAE encoder to produce another latent representation, also with a channel dimension of 4.

The mask generated by SAM, which identifies the specific region of interest (e.g., a T-shirt), is resized to match the spatial dimensions of the latent representations. Although the mask does not undergo encoding, it plays a crucial role in guiding the integration process. The resized mask, with a single-channel dimension, is concatenated with the latent representations from the DragGAN-modified image and its masked-out counterpart. This concatenation results in a unified latent representation with a channel dimension of 9.

### 13.2.3.2   Guided Denoising for Precision and Detail

The concatenated latent representation, along with user-provided textual descriptions of the desired clothing attributes (e.g., "A young woman stands on a quiet street, exuding confidence in her bright orange T-shirt adorned with playful cartoon characters."), is fed into the SDXL inpainting pipeline. The textual input serves as a conditioning factor, enabling the model to integrate semantic information about the desired clothing style directly into the denoising process.

The inpainting process proceeds iteratively, guided by a denoising schedule that progressively refines the latent representation. At each step, the SDXL model applies

TABLE 13.1    SDXL-VAE's reconstruction performance on the COCO2017 validation split, images of size 256*256 pixels. It outperforms the original model in all metrics.

| Model | PSNR ↑ | SSIM ↑ | LPIPS ↓ | rFID ↓ |
|---|---|---|---|---|
| *SDXL-VAE* | **24.7** | **0.73** | **0.88** | **4.4** |
| *SD-VAE 1.x* | 23.4 | 0.69 | 0.96 | 5.0 |
| *SD-VAE 2.x* | 24.5 | 0.71 | 0.92 | 4.7 |

its denoising diffusion network, which incorporates the textual conditioning and latent-space information to align the generated image with user specifications. This iterative refinement ensures that the output is both visually coherent and semantically accurate, maintaining high fidelity to the user's requested modifications.

### 13.2.3.3    *Decoding the Refined Latents into Visual Outputs*

Once the iterative denoising process is complete, the final latent representation, now fully refined, is passed through the VAE decoder of the SDXL model. The decoder reconstructs a high-resolution image from the latent representation, translating the compact feature map back into a detailed visual output. The result is a highly customized image that reflects the modifications specified by the user, including both the structural changes introduced by DragGAN and the stylistic details guided by textual inputs.

This detailed integration of VAE-based latent representations and diffusion-based inpainting showcases the flexibility and power of SDXL as a generative tool for virtual clothing modification. The SDXL-VAE's reconstruction performance on the COCO2017 validation split is shown in Table 13.1, the SDXL-VAE outperforms original model in all evaluated reconstruction metrics [169].

By strategically combining data from multiple sources—DragGAN-modified image and its masked counterpart, segmentation mask, and user description—the framework achieves a level of personalization and visual quality that surpasses traditional image editing techniques. Each module contributes to the final output in a specialized way, ensuring that both the global structure and fine-grained details of the garment can be customized to meet user specifications. This approach opens new possibilities for creative exploration in fashion design and beyond.

## 13.3    EXPERIMENTS

To validate the effectiveness of the proposed method, we conduct a series of experiments in the context of virtual clothing design.

### 13.3.1    Dataset Collection and Specifications

#### 13.3.1.1    *Scope and Objectives*

The dataset serves as the cornerstone for developing a comprehensive virtual clothing system. To meet the diverse requirements of garment types, seasonal styles, and user

**Figure 13.5** A glimpse of our collected datasets. The dataset showcased in the glimpse image serves as a comprehensive resource for virtual clothing customization, containing over 10 million images that represent a wide range of clothing categories.

preferences, the data collection process was designed to ensure extensive coverage and high quality. The dataset contains more than 10 million images to comprehensively represent various clothing categories. A glimpse can be seen as Figure 13.5.

The dataset showcased in the Figure 13.5 serves as a comprehensive resource for virtual clothing customization, containing over 10 million images that represent a wide range of clothing categories. It is meticulously organized into subcategories for both Men's and Women's Wear, covering various types, styles, materials, and seasonal suitability. The dataset is designed to support advanced AI-driven fashion design and modification tasks, ensuring high-quality, detailed, and diverse garment representations. Its strict quality standards and balanced distribution of poses and angles make it an ideal foundation for training models to handle complex virtual clothing modifications with precision and realism.

This dataset is specifically curated to train the SDXL-inpainting model for virtual clothing customization. It includes over 10 million high-resolution images, meticulously categorized by garment type, style, and material, ensuring diverse and detailed representations. The dataset's focus on clarity, quality, and variety enables the SDXL model to learn precise inpainting techniques, allowing it to generate realistic and customized clothing designs based on user inputs. Its robust structure supports the model's ability to handle complex modifications, such as texture, pattern, and color changes, with high fidelity.

## 13.3.1.2 *SDXL's Training*

The dataset plays a crucial role in training a specialized SDXL-inpainting model tailored for virtual clothing customization. The dataset's extensive coverage of diverse garment types, styles, and materials provides a rich source of data for the model to learn from. To ensure optimal training, the dataset undergoes several pre-processing steps:

- Random Mask Generation: Instead of relying on manual segmentation or automated techniques, we adopt a random approach to mask generation. A random polygon region is selected within each image, encompassing various areas of the garment. This randomization introduces diversity and complexity, challenging the model to inpaint a wide range of scenarios and learn to adapt to different garment structures and details.

- Textual Descriptions: Each image in the dataset is annotated with a textual description that specifies the desired attributes of the original, unmasked garment. These descriptions provide valuable semantic information that guides the model in generating inpainted images that align with user preferences for the garment's overall appearance, including color, texture, pattern, and style.

The training process involves iteratively fine-tuning the SDXL-inpainting model using the preprocessed dataset. The model architecture is optimized to generate realistic and high-quality inpainted images. The textual descriptions guide the model to learn and incorporate semantic information, further improving the accuracy and personalization of the inpainting results.

By leveraging this carefully curated dataset and training a specialized SDXL-inpainting model with randomly generated masks, we can achieve great results in virtual clothing customization. The trained model can then be used to generate customized clothing designs based on user input, offering a powerful tool for fashion designers, e-commerce platforms, and individuals seeking personalized clothing options.

## 13.3.1.3 *Dataset Composition*

The dataset is divided into two primary categories: Men's Wear and Women's Wear, each further classified into subcategories and detailed classifications.

For Men's Wear, the primary subcategories include upper wear, trousers, and formal suits. Examples of upper wear are T-shirts, shirts, and jackets, while trousers include jeans, chinos, and formal pants. Formal suits are further subdivided into business suits, tuxedos, and blazers. For Women's Wear, subcategories include outerwear, upper wear, dresses, and trousers. Outerwear includes items such as coats and cardigans, while upper wear consists of blouses, shirts, and tops. Dresses range from casual to evening gowns, and trousers encompass options like leggings and formal pants.

Each subcategory is further detailed based on specific attributes, such as type, seasonal suitability, style, fit, material, thickness, collar type, functional features, fastening style, garment length, sleeve length, pattern, and fabric techniques. For example:

- Type: Differentiates between polo shirts, hoodies, trench coats, and other garment types.

- Seasonal Suitability: Indicates whether the clothing is for spring, summer, autumn, or winter.

- Style and Fit: Categorizes garments into casual, formal, slim-fit, oversized, and more.

- Material and Thickness: Specifies fabrics like cotton, polyester, or silk, and their relative weights.

- Functional Features: Includes waterproof, thermal, or UV-protective properties.

- Design Elements: Incorporates collar types (e.g., V-neck, turtleneck), patterns (e.g., floral, geometric), and techniques (e.g., knitted, embroidered).

This multi-layered classification ensures that the dataset is highly detailed and adaptable to diverse use cases, providing a rich resource for training AI systems capable of handling complex modifications.

### 13.3.1.4 Image Quality and Presentation Standards

The dataset adheres to strict quality standards to ensure its effectiveness for virtual clothing applications. All images meet the following criteria:

- Resolution and Aspect Ratio: Each image has a minimum resolution of 1024 × 1024 pixels to ensure sufficient detail. There are no restrictions on aspect ratios, providing flexibility in image composition.

- Garment Visibility: Clothing items in the images must be clearly visible, free from occlusions caused by accessories, body parts, or other objects. Each image focuses on a single clothing item, either standalone or worn by a model.

- Garment Presentation Styles: Clothing can be displayed on mannequins, laid flat, or worn by models. For model-based images, the garment should remain the focal point, with minimal distractions such as elaborate poses or accessories.

- Background Simplification: Simple or solid-colored backgrounds are preferred to ensure that the clothing remains the central focus. In cases where pure backgrounds are not feasible, visually simplified environments were chosen.

- Fashion Relevance: The dataset prioritizes modern and contemporary styles that align with current fashion trends to maintain applicability in real-world scenarios.

### 13.3.1.5 *Diversity in Angles and Poses*

To enhance robustness, the dataset includes a balanced distribution of poses:

- Over 70% of the images feature front-facing views to provide maximum clarity of garment details.

- Approximately 25% of the images are side profiles, captured at angles of less than 30 degrees.

- The remaining 5% include rear views or other creative poses to increase diversity.

The dataset ensures that all subcategories, as well as their detailed classifications, are evenly represented. This balance prevents over-representation of any single type, enabling a wide range of virtual clothing modifications.

### 13.3.2  Shape Modification and Masking

In the first experiment, users modify the shape of a garment (e.g., a T-shirt) using DragGAN. After the shape transformation, SAM is applied to generate the corresponding mask. The results show that SAM effectively isolates the garment area, providing clean segmentation with minimal manual intervention. A showcase of Drag-GAN and SAM, shape modification and masking, can be seen in the Figure 13.6, where the result shows that SAM effectively isolates the garment area, providing clean segmentation with minimal manual intervention:

The Figure 13.6 showcases the effectiveness of combining DragGAN and SAM for precise clothing modifications and segmentation. In the first step, DragGAN is utilized to reshape the sleeves of the T-shirt by dragging control points to the desired positions, enabling intuitive and detailed adjustments to the garment's structure.

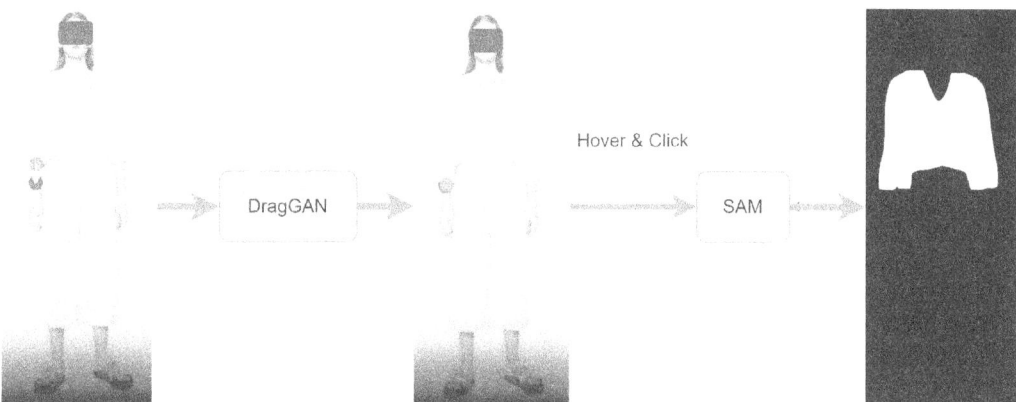

Figure 13.6   The showcase of DragGAN and SAM, shape modification and masking. The result shows that SAM effectively isolates the garment area, providing clean segmentation with minimal manual intervention.

Following this, SAM accurately identifies and segments the modified T-shirt, generating a clear and precise mask of the clothing area. This workflow not only demonstrates the seamless integration of DragGAN's shape-editing capabilities with SAM's robust segmentation performance but also highlights the potential for high-quality customization and further processing in virtual clothing modification tasks.

The integration of DragGAN and SAM exemplifies a powerful workflow where structural modifications and precise segmentation are seamlessly connected, paving the way for high-quality, user-driven virtual clothing customization.

### 13.3.3   Inpainting for Detail Customization

In the second experiment, users provide specific design preferences for the garment. These preferences are input into the SDXL inpainting model along with the masked image and the original mask. The results demonstrate that SDXL can successfully generate realistic images, meeting user expectations for clothing customization.

The testing of SDXL's inpainting functionality, shown in Figure 13.7, highlights its remarkable ability to generate visually compelling results that adhere closely to user-provided textual descriptions. Using the provided inputs—namely, the original image, the segmented mask of the T-shirt, and the masked image with the T-shirt area removed—the inpainting model successfully applied the specified modifications. Guided by the description "Red color, cartoon pattern in the center," SDXL transformed the T-shirt into a vibrant red garment featuring a clearly defined cartoon pattern prominently displayed in the center. The outcome demonstrates SDXL's capacity

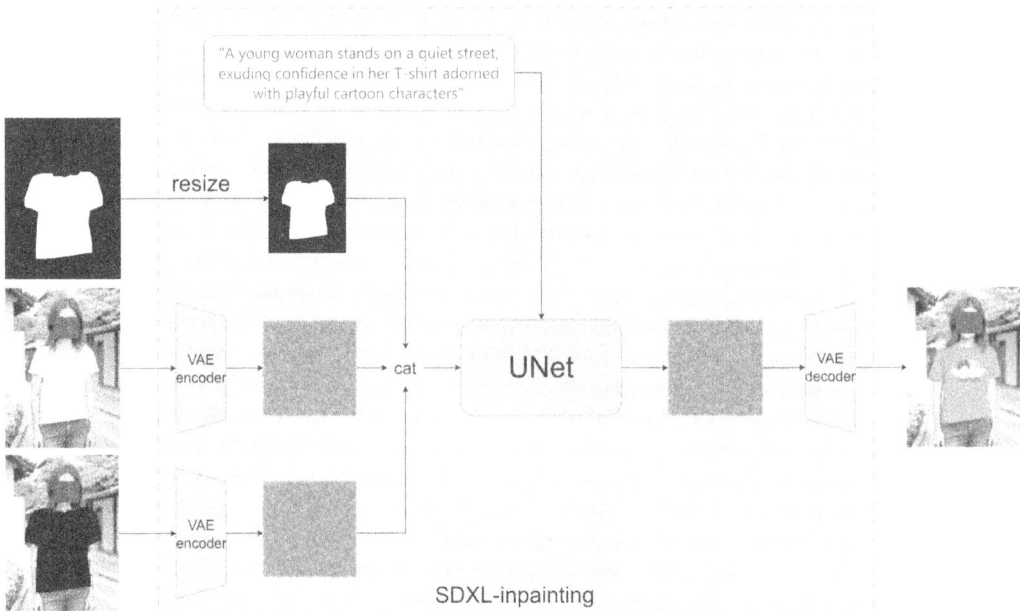

Figure 13.7   The showcase of SDXL's inpainting for detail customization. The results demonstrate that SDXL can successfully generate realistic images, meeting user expectations for clothing customization.

to interpret and translate high-level textual guidance into precise visual alterations, ensuring that the generated content aligns seamlessly with user expectations.

This result underscores the robustness of the SDXL inpainting pipeline, where it leverages the combination of visual and semantic inputs to deliver coherent and contextually accurate outputs. The integration of the mask ensures that changes are localized to the intended region, preserving the surrounding areas of the image without unwanted artifacts. Furthermore, the model's ability to render fine-grained details, such as the cartoon pattern, while maintaining a natural and realistic appearance of the clothing, highlights its potential for creative design tasks. By effectively synthesizing elements of color, texture, and pattern based on user-defined descriptions, SDXL demonstrates its versatility and power as a tool for advanced virtual clothing customization.

### 13.3.4 User Satisfaction Study

A user satisfaction study is conducted to evaluate the overall experience with the system. Participants are asked to rate their satisfaction based on ease of use, the quality of garment modification, and the level of customization.

The result is shown in Figure 13.8, the majority of scores cluster around the higher range, specifically between 7 and 9, with a peak at scores of 8 and 9. This indicates a

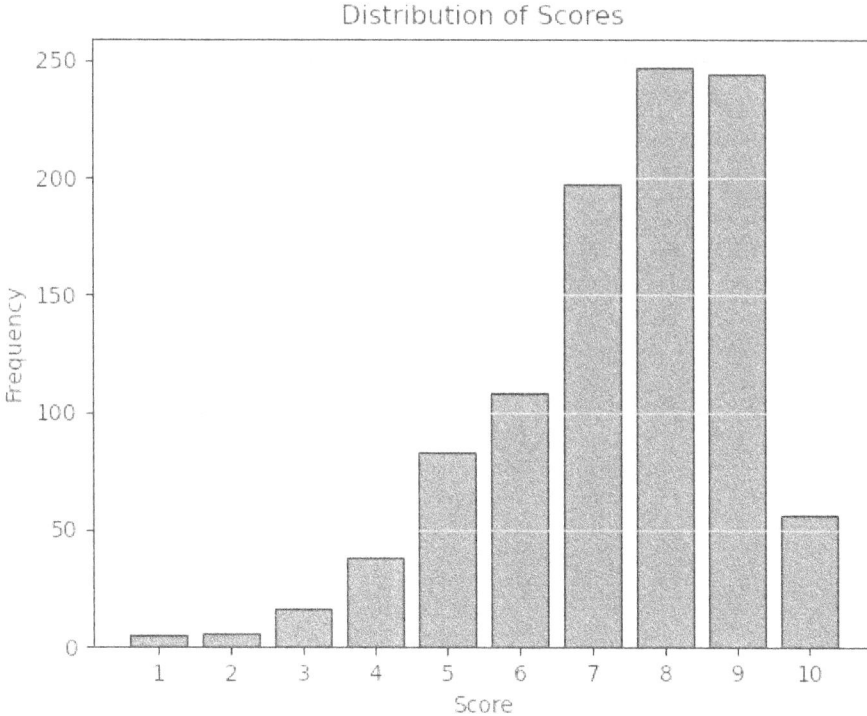

Figure 13.8 The distribution of scores.

strong preference and satisfaction among users who interacted with the system. The provided histogram clearly illustrates the high level of user satisfaction achieved by the proposed algorithm. A significant majority of scores are concentrated in the 7–9 range, with peaks at 8 and 9, suggesting that users consistently found the system effective and aligned with their expectations. The sparse occurrence of lower scores, particularly between 1 and 4, highlights the algorithm's robustness and reliability in delivering satisfactory results.

This strong performance can be attributed to the seamless integration of advanced components such as DragGAN for structural modifications, SAM for precise segmentation, and SDXL for detailed inpainting and customization. Each component contributes uniquely to the overall workflow, ensuring that the outputs not only meet but often exceed user expectations. Moreover, the algorithm's ability to incorporate user-defined inputs, such as textual descriptions, allows for a high degree of personalization, which is likely a key factor in the overwhelmingly positive reception.

The distribution of satisfaction scores underscores the algorithm's ability to handle complex tasks effectively, maintaining consistency and quality across diverse scenarios. By combining technical precision with user-centered design, the system demonstrates its capacity to address nuanced requirements, setting a new benchmark for virtual clothing modification. The histogram serves as clear evidence of the algorithm's success in translating advanced generative capabilities into practical and user-friendly applications, confirming that the combined approach offers a compelling solution for virtual clothing design.

## 13.4 CONCLUSION

This work presents a novel image editing paradigm specifically for clothing fashion customization that combines DragGAN for shape manipulation with advanced semantic segmentation and inpainting techniques using SAM and SDXL. The system provides users with an intuitive and flexible platform for both reshaping garments and altering their finer details, such as color, texture, and pattern. The experimental results demonstrate the effectiveness of the method in real-world scenarios, where both technical accuracy and user satisfaction are paramount. The proposed methodology has significant implications for various industries, including fashion design, virtual try-ons, and digital content creation. Future work will focus on further improving the model's capabilities, exploring real-time applications, and expanding the system to support a wider range of garment types and customizations.

# CHAPTER 14

# Retrieval-Augmented Generation Paradigm for Professional Knowledge Acquisition Applications

## 14.1 INTRODUCTION

With the continuous development of artificial intelligence technology, generative AI has demonstrated exceptional performance in various tasks, such as the application of large language models (LLMs) in text [34, 217] and code generation [95, 247], as well as the emergence of image [186] and video generation models [141]. However, these generative models still face several challenges, including outdated knowledge, difficulty in processing long-tail data, vulnerability to private training data leakage, and high costs for both training and inference.

Fine-tuning techniques effectively enhance the timeliness of model knowledge and its ability to handle long-tail data by conducting additional training on pre-trained generative AI models using the latest and targeted data specific to particular tasks or domains. However, fine-tuning also incurs significant costs and time consumption, limiting its application in scenarios requiring frequent updates and potentially introducing risks of privacy breaches. Furthermore, while fine-tuning can enhance model performance on specific tasks, it does not fully address the issue of insufficient generalization capability. Therefore, although fine-tuning alleviates some challenges, it also brings limitations in terms of efficiency, privacy, and generalization.

To address the aforementioned issues, we propose a retrieval-augmented generation paradigm for professional knowledge acquisition applications, aiming to combine the model's generative capability with information retrieval technology to achieve accurate and real-time text generation. Retrieval-augmented generation integrates language models with information retrieval techniques by enhancing the generation process through retrieving relevant information from external knowledge bases. Specifically, when the model needs to answer a question or generate content, it first

DOI: 10.1201/9781003644972-14

retrieves relevant information from external knowledge bases and then integrates this information as input to assist the model in producing more accurate and relevant outputs.

Our proposed paradigm is versatile and can be widely applied in various industries, including enterprise, healthcare, education, etc. We validate the reliability of our proposed paradigm using three scenarios: equipment fault diagnosis, hospital inquiry and university admissions.

## 14.2 METHODOLOGY

Our proposed paradigm primarily consists of three core steps: data preparation, query retrieval, and large model generation. The detailed processing flow is illustrated in Figure 14.1.

### 14.2.1 Data Preparation

In the data preparation phase, constructing an extensive and deep knowledge base is crucial. This process involves collecting documents from various sources, including but not limited to books, academic papers, business manuals, technical documents, policies, and regulations, through means such as downloads from open-source websites and API interfaces. For unstructured or image-based data, techniques such as Optical Character Recognition (OCR) are applied to convert them into editable text formats and segment them into smaller, more manageable text chunks to enhance efficiency. To enable semantic-based information retrieval, pre-trained language models such as

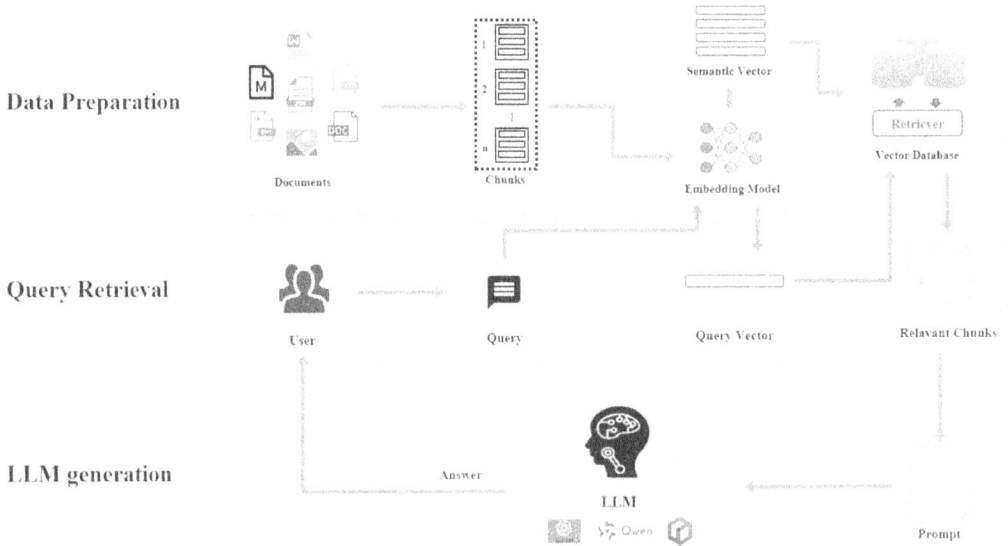

Figure 14.1 The processing flow of the proposed retrieval-augmented generation paradigm.

BERT [44] and RoBERTa [140] are utilized as embedding models to encode these text chunks, transforming them into semantic vector representations in a high-dimensional space. Finally, all processed text vectors, along with their corresponding metadata, are indexed and stored in an efficient vector database such as Faiss [49] or Milvus [227], ensuring rapid query response times.

## 14.2.2 Query Retrieval

During the query retrieval phase, when a user poses a question, the first step involves converting the user's natural language query into a corresponding query vector representation using the same embedding model as that of the data preparation phase. Subsequently, approximate near-neighbor search (ANN search) or other similarity searching methods are performed within the vector database to identify a set of document chunks that are most similar to the query vector. This step relies on metric methods such as cosine similarity and Euclidean distance to determine the best matches.

## 14.2.3 LLM Generation

The retrieved relevant document chunks are integrated into a prompt template as an external knowledge source, forming an input sequence that includes contextual information. This sequence is then passed to a LLM, such as the GPT series [16, 159, 174, 175] or other advanced pre-trained models. The LLM leverages its robust generalization capabilities and language understanding capabilities to fully consider the information retrieved along with its own learned knowledge, generating a more accurate and contextually appropriate response. During this process, decoding strategies, such as sampling or beam search, can be introduced to improve the quality of the output.

Through these three stages, the retrieval-augmented generation paradigm effectively utilizes external knowledge, enhancing the performance of natural language processing tasks, especially those requiring timely and precise factual information retrieval. This approach not only improves the accuracy of the system's responses, but also enables it to continuously update its knowledge base, staying current with the latest information.

## 14.3 EXPERIMENTS AND RESULTS

### 14.3.1 Application in Equipment Fault Diagnosis Scenarios

Traditional equipment fault diagnosis has largely depended on manual experience and periodic inspections, a method that is not only inefficient but also overly reliant on human resources. Furthermore, as equipment technologies continuously evolve, traditional diagnostic approaches may fail to promptly adapt to new fault patterns and maintenance requirements, resulting in delayed knowledge updates.

To address these challenges, we integrate retrieval-augmented generation technology into the equipment fault diagnosis assistant. By consolidating a multi-source

Query: What could be the possible causes for fault code C113017?
Answer: Possible causes of fault C113017 include wiring faults, left EPB
caliper faults, and internal ECU faults in the left drive circuit.
References:
- x System Repair Manual.pdf
- xx System Repair Manual.pdf
- xxx Repair Manual.pdf

Figure 14.2   An application example in equipment fault diagnosis scenarios.

knowledge system encompassing maintenance manuals, fault cases, engineering documents, and more, we enable functionalities such as natural language-based fault
cause analysis, fault information retrieval, and intelligent diagnostic assistance. This
integration not only enhances the accuracy and efficiency of diagnostics but also reduces dependency on technical personnel. As demonstrated in an application example
shown in Figure 14.2, when a technician inputs a query regarding the possible causes
of fault code C113017, the fault diagnosis assistant can provide precise cause analysis
by leveraging an externally attached domain-specific knowledge base.

## 14.3.2   Application in Hospital Inquiry Scenarios

In traditional medical information inquiries, healthcare professionals rely on static
knowledge bases, paper manuals, and electronic documents. These resources are characterized by slow update cycles and are difficult to maintain up-to-date, resulting in
information lag. Confronted with complex medical policies and evolving operational
guidelines, healthcare providers must spend considerable time cross-referencing multiple sources to ensure accuracy, a process that is inefficient and increases the risk
of human error, which is especially critical during emergencies, potentially impacting
patient care outcomes. Moreover, the fragmentation of information poses significant
challenges; varying procedural differences across departments cannot be effectively
consolidated by traditional systems, complicating interdepartmental collaboration
and making it particularly difficult for new doctors and nurses to rapidly locate essential materials.

By integrating retrieval-augmented generation technology into intelligent medical
consultation assistants, the aforementioned issues can be alleviated. Firstly, retrieval-
augmented generation facilitates the construction of a dynamic knowledge base updated in real time, ensuring all information is timely and accurate. Healthcare professionals can consult the latest interpretations of medical policies, clinical guidelines,
and various operational procedures within the hospital without worrying about outdated or incomplete data. Secondly, retrieval-augmented generation achieves comprehensive integration of information, covering broad medical fields while it cus-
tomizes answer content according to specific scenarios and user needs, ensuring provided information is both precise and practical. As illustrated in Figure 14.3, when
a doctor queries "How to apply for medical insurance?," the retrieval-augmented
generation medical assistant returns an accurate response including the most recent

Query: How to apply for medical insurance?
Answer: According to the information in the knowledge base, the process for applying for medical insurance is as follows:

1. Selection of a designated medical institution for prenatal examination:
- The insured should submit the required application materials to the medical insurance office to complete the procedures for selecting a designated medical institution for prenatal examination under the maternity insurance program.
- The medical insurance office will review the application materials and enter them into the maternity insurance information system.
- After stamping, the "Designated Medical Institution Form for Prenatal Examination under Employee Maternity Insurance" will be returned to the insured.
- Special Note: If the insured selects "xx Hospital" as the designated medical institution for prenatal examination, they can only enjoy the corresponding benefits at the xx of the hospital; if "xx Hospital" is selected, benefits can only be enjoyed at the xx campus.
- Alternatively, log in to the "xx Government Service Portal" at https://xxx, search for "Designated Prenatal Examination," and follow the instructions to apply.
2. Selection of a designated medical institution for general outpatient (emergency) services:
- For employee medical insurance, the insured can directly select a hospital as the "larger" medical institution; for resident medical insurance, the insured must first select one primary medical institution ("smaller") before selecting a hospital ("larger").
- Offline registration: The insured should visit the registration desk with their social security card, electronic medical insurance voucher, or resident ID card to complete the process.
- Online registration includes steps such as scanning the Guangzhou medical insurance outpatient selection QR code and registering for medical insurance in another city within the province.
3. Management of admission and discharge for work-related injury patients:
- The designated medical institution should promptly handle the discharge procedures for work-related injury patients who meet the discharge criteria and must not deliberately extend their hospital stay.
- If the patient delays or refuses to be discharged, all expenses incurred will be borne by the patient themselves.
4. Registration for medical treatment in another city within the province or across the province:
- Registration for medical treatment in another city within the province can be handled through the "xx Medical Insurance" mini-program or by calling the local social security bureau.
- Registration for medical treatment across the province can be handled through the "National Medical Insurance Registration for Cross-province Medical Treatment" mini-program or the "National Medical Insurance Service Platform" app.

References:
- Guide to designated medical insurance for xxx Hospital.pdf

**Figure 14.3** An application example in hospital inquiry scenarios.

methods for processing designated medical insurance along with reference documents for traceability.

## 14.3.3 Application in University Admissions Scenarios

During the admissions phase, universities face significant challenges in information management and response efficiency. Traditional consultation methods primarily depend on manual efforts, with staff responding individually to student and parent inquiries via phone, email, or online chat tools. This approach not only consumes substantial human resources but also struggles to guarantee the speed and accuracy of responses. Especially during peak admission periods, the surge in consultation volume leads to severe delays in replies, sometimes taking several days to receive an answer, significantly affecting students' decision-making processes and experiences.

The introduction of a retrieval-augmented generation-based intelligent consultation assistant tailored for higher education institutions can provide rapid and accurate

Query: What are the admission consultation services and contact details for xx University?
Answer: The enrollment consultation telephone number of xx University includes: 09xx-xxxxxx1, 09xx-xxxxxx2, 09xx-xxxxxx3 and 09xx-xxxxxx4.
References:
- xx University basic information.pdf

Figure 14.4   An application example in university admissions scenarios.

responses to inquiries from students and parents, which greatly enhances user experience and service quality. The system can instantly respond to high volumes of consultations, ensuring quick information access regardless of whether it's peak or regular times. It integrates up-to-date information from multiple channels ensuring consistent and transparent information across the university, providing authoritative and detailed answers and reducing human errors. It automates handling common questions, lessening the burden on staff and continuously updates the knowledge base based on feedback. As illustrated in Figure 14.4, when asked 'What are the admission consultation services and contact details for xx University?' the assistant accurately displays its contact information.

## 14.4   CONCLUSION

This chapter delves into an innovative retrieval-augmented generation paradigm that ingeniously combines the strengths of information retrieval with deep learning-based generative models, offering a robust new solution for applications within the artificial intelligence industry. By integrating precise information retrieval mechanisms with advanced language generation capabilities, retrieval-augmented generation not only significantly enhances the accuracy and reliability of query responses but also effectively reduces reliance on large-scale annotated data, thereby decreasing the cost and complexity associated with model training.

As technology continues to evolve and application domains expand, retrieval-augmented generation technology is demonstrating immense potential. It is poised to play a pivotal role in various fields such as intelligent customer service, automated question-answering systems, personalized recommendations, and beyond, driving advancements across these industries. Looking ahead, we can anticipate that retrieval-augmented generation will leverage its unique advantages in an even broader range of tasks, further enhancing human-computer interaction experiences. Additionally, it provides a solid foundation for researchers to explore new frontiers in artificial intelligence, fostering innovation and discovery.

In summary, the retrieval-augmented generation paradigm represents a significant leap forward in AI-driven solutions, promising not only to improve current applications but also to open up new possibilities for future development.

# CHAPTER 15

# Multimodal Generation Paradigm for Visualizing Historical Artifacts

## 15.1 INTRODUCTION

The advent of text-to-image generation models has opened new opportunities in cultural heritage preservation and promotion. This monograph introduces an innovative paradigm combining text-to-image [243] models with an image prompt adapter (IPAdapter) [242] to transform descriptions and images of historical artifacts into corresponding human character designs. This paradigm is poised to enhance cultural education, entertainment, and public engagement by visualizing history in ways previously unattainable. Our method employs fine-tuned Low-Rank Adaptation (LoRA) [84] models to generate desired art styles and IPAdapter for seamless stylistic consistency between the artifact image and the generated human character. Through qualitative examples, we demonstrate the potential of this approach to revolutionize cultural storytelling and education.

Cultural heritage is a tangible link to our past, preserving the stories, traditions, and values of civilizations. It serves as a repository of human knowledge and creativity, fostering cultural identity, understanding, and appreciation. However, traditional methods of engaging with cultural heritage often lack interactivity and fail to capture the imagination of modern audiences. The advent of generative AI, particularly Text-to-Image models, has opened up new avenues for creative expression and content generation. These models can generate realistic images based on textual descriptions, offering unprecedented possibilities for storytelling, art, and design. By harnessing the power of AI with the rich tapestry of historical artifacts, we can breathe new life into cultural heritage, making it more accessible and engaging.

This chapter introduces a novel paradigm that combines Text-to-Image models with IPAdapter techniques to create lifelike character images based on cultural relics and designed descriptions. This approach allows users to imagine and visualize the people and personalities behind historical artifacts, fostering a deeper connection with the past.

DOI: 10.1201/9781003644972-15

The applications of this paradigm are diverse and far-reaching. It can be utilized in various fields, including Education, Entertainment, Cultural Tourism, and Art and Design.

## 15.2 METHODOLOGY

Our proposed paradigm combines text-to-image generation models with IPAdapter to achieve seamless alignment between the visual characteristics of cultural artifacts and the generated human character designs. Below, we detail the technologies and techniques underpinning this system.

### 15.2.1 Text-to-Image Generation

Text-to-image generation is a rapidly evolving field powered by advanced deep learning models, such as diffusion models [38]. These models translate textual prompts into highly detailed images by iteratively denoising a random noise input, guided by the semantic and stylistic information encoded in the text. Diffusion models are generative models trained to reverse a noise corruption process, reconstructing images from random noise. During inference, the model iteratively applies learned denoising steps, guided by textual embeddings, to generate coherent and high-quality images. LoRA is a parameter-efficient fine-tuning method designed to adapt pre-trained large-scale models for specific tasks or styles. In our monograph, we utilize LoRA to fine-tune the model efficiently, enabling it to adapt to specific cultural and artistic styles while maintaining computational efficiency. Instead of retraining the entire model, LoRA introduces low-rank matrices that modify only a small subset of the model's parameters during training. In this chapter, LoRA guides the model to focus on generating images within the desired artistic style, ensuring consistency and precision while maintaining computational efficiency.

### 15.2.2 Image Prompt Adapter (IPAdapter)

The IPAdapter is a crucial component of our framework, responsible for transferring the stylistic and aesthetic elements of cultural artifacts into the generated human character designs.

The IPAdapter operates as a feature extraction and transformation module, ensuring that the generated outputs align visually with the provided artifact image. The IPAdapter leverages a pre-trained vision model (e.g., a convolutional neural network [36] or Vision Transformer [107]) to extract high-level features from the artifact image. These features capture details like color palette, texture, shape, and patterns, which are critical for stylistic consistency. Extracted features are embedded into the latent space of the text-to-image generation process. This is achieved by conditioning intermediate layers of the generative model on the artifact's features, ensuring that the final output reflects both the textual prompt and the visual essence of the artifact.

By directly incorporating visual elements from the artifact, the IPAdapter ensures that the generated characters are not just visually appealing but also culturally

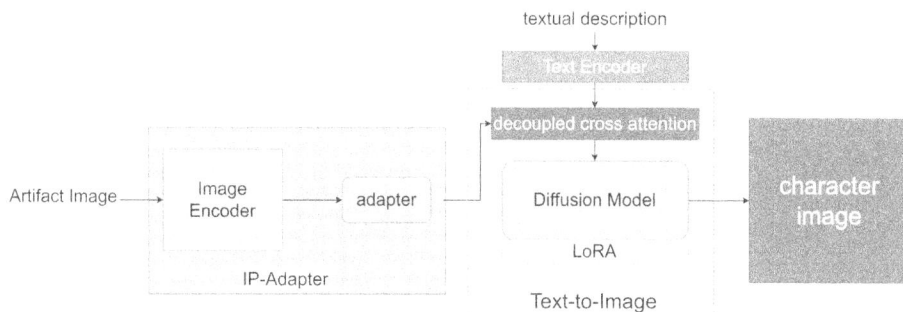

Figure 15.1    Workflow of our paradigm.

and aesthetically coherent. The module can adapt to various artifact types, enabling wide applicability across different cultural domains. For each input artifact image, the IPAdapter extracts stylistic attributes and infuses them into the text-to-image generation pipeline. This ensures that the final output maintains a high degree of visual harmony between the artifact and the generated human character.

## 15.2.3   Workflow

The workflow of our paradigm is designed to effectively combine textual descriptions and artifact images to generate visually coherent human character designs.

As shown in Figure 15.1, the process begins with the user providing an image of a cultural artifact as input. This artifact image contains the stylistic and aesthetic attributes to be transferred to the generated human character. The artifact image is passed through the IPAdapter, which extracts its high-level visual features, such as texture, color patterns, and stylistic nuances. Specifically, the artifact image is passed through an Image Encoder. This encoder extracts latent visual features that represent the artifact's texture, shape, and stylistic elements, which will influence the downstream generation process. The extracted features are then processed through an adapter module, specifically designed to align the image encoder's output with the required input format of the main model. This ensures smooth integration of visual features into the system while maintaining computational efficiency. Alongside the artifact image, a descriptive textual prompt is provided by the user. This text, processed through a Text Encoder, serves to define the semantic content of the human character, such as their role, personality, or historical context. A decoupled cross-attention module is used to integrate the artifact features (from the adapter) and the textual description (from the text encoder). By decoupling these inputs, the model can independently weigh the importance of artifact-inspired visual cues and textual directives, enabling a balanced fusion. LoRA ensures that the model is capable of producing outputs in the desired artistic style. The extracted features from the IPAdapter are integrated into this generation process to achieve stylistic alignment with the artifact image. Then the model generates high-quality images of human characters. These images are a fusion of the semantic details described in the text and the stylistic elements extracted from the artifact image. The result is a human character that not only aligns visually with the artifact but also embodies the narrative conveyed by the text.

By following this workflow, the system enables seamless interaction between visual and textual modalities, allowing users to create culturally and artistically meaningful character designs with ease.

## 15.3   RESULTS AND EXAMPLES

Two representative examples are provided below:

### 15.3.1   Example 1

Figure 15.2 highlights the remarkable results achieved through our proposed paradigm and model. The inputs include:

- A visual input: the Shang and Zhou Dynasties Gold Ornament Depicting the Solar Bird, a historically significant artifact with intricate details and symbolic design elements.

- A textual description: Archer. Her jade-like face was gentle and radiant, her eyes bright and clear. Her delicate wings fluttered gracefully, their golden light shimmering and dazzling, akin to the essence of the sun itself.

Leveraging the unique strengths of our paradigm, which integrates fine-grained visual conditioning with textual semantics, and our customized model architecture, the output is a breathtaking representation of a female archer. The generated image masterfully incorporates the artistic elements of the solar bird ornament, blending them harmoniously with the narrative details from the text. The archer's golden wings, radiant attire, and serene expression embody the elegance and symbolism inherent in the input.

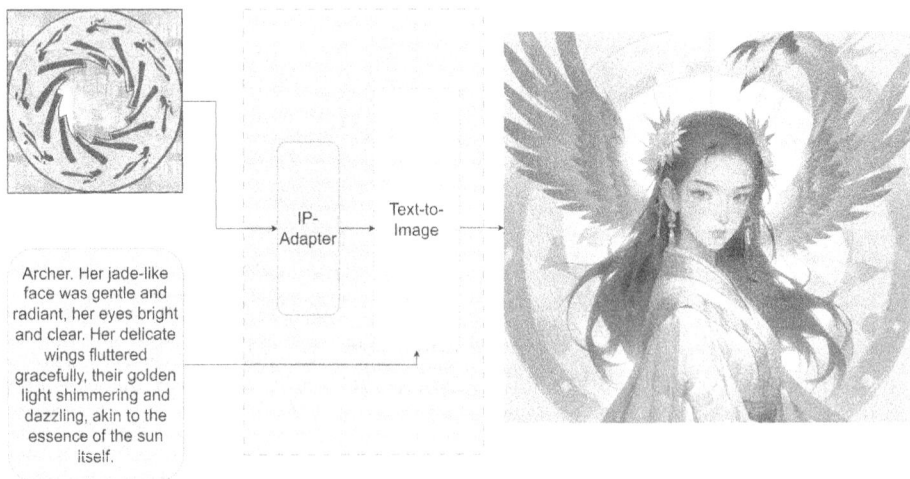

Figure 15.2   Example of the Shang and Zhou Dynasties Gold Ornament Depicting the Solar Bird's generated result.

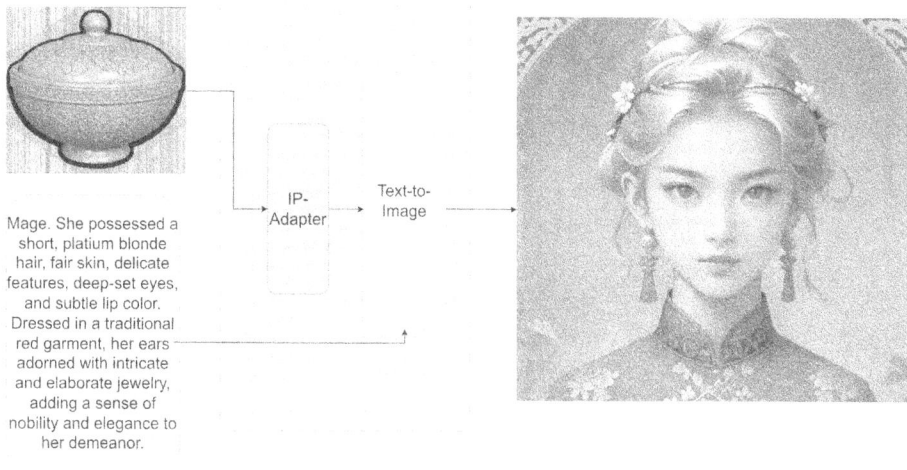

Mage. She possessed a short, platium blonde hair, fair skin, delicate features, deep-set eyes, and subtle lip color. Dressed in a traditional red garment, her ears adorned with intricate and elaborate jewelry, adding a sense of nobility and elegance to her demeanor.

IP-Adapter

Text-to-Image

**Figure 15.3** Example of the Yongle Marked Carved Red Lacquer Bowl with Flower Patterns's generated result.

This result exemplifies the model's capability to seamlessly merge cultural aesthetics with imaginative storytelling, delivering visually stunning and contextually rich outputs. It is a testament to the paradigm's ability to produce art that resonates deeply with both historical inspiration and creative vision.

## 15.3.2 Example 2

Figure 15.3 showcases the excellent performance of our paradigm and model. The inputs include:

- A visual input: the Yongle Marked Carved Red Lacquer Bowl with Flower Patterns, a historically significant artifact known for its intricate floral carvings and vibrant red lacquer finish.

- A textual description: Mage. She possessed short platinum blonde hair, fair skin, delicate features, deep-set eyes, and subtle lip color. Dressed in a traditional red garment, her ears adorned with intricate and elaborate jewelry, adding a sense of nobility and elegance to her demeanor.

Our paradigm combines fine-grained visual conditioning with semantic understanding from the textual description. This enables the output: a stunning depiction of a mage. The model skillfully integrates the aesthetic and ornate qualities of the lacquer bowl into the mage's design. Her traditional red garment reflects the bowl's vibrant red tone, while the floral jewelry echoes the intricate patterns of the artifact. The character's noble and graceful demeanor aligns perfectly with the artifact's historic elegance.

The result highlights the model's ability to synthesize cultural elements and descriptive narratives into visually harmonious and artistically rich outputs. It is a testament to the paradigm's sophistication and creative potential.

Both examples demonstrate the model's ability to capture the essence of cultural artifacts while maintaining visual and stylistic coherence. Our paradigm excels at seamlessly integrating visual and textual inputs, transforming cultural artifacts and descriptive narratives into visually stunning and artistically harmonious outputs. This showcases the model's exceptional creativity and its ability to preserve both aesthetic and contextual coherence.

## 15.4  CONCLUSION

In this chapter, we proposed an innovative paradigm that integrates multimodal inputs, combining cultural artifacts with textual descriptions to generate visually compelling and contextually accurate images. By leveraging Text-to-Image models and IPAdapter techniques, the approach demonstrated remarkable capabilities in preserving cultural essence while enhancing artistic expression. Through the effectiveness of our paradigm, verified by the showcase results, we highlighted its ability to transform artifacts such as the Shang and Zhou Dynasties Gold Ornament Depicting the Solar Bird and the Yongle Marked Carved Red Lacquer Bowl with Flower Patterns into visually captivating outputs. These results underline the system's capacity for aesthetic refinement and contextual coherence, affirming its potential in broader applications such as cultural heritage preservation, digital art generation, and creative industries. This paradigm not only offers a practical framework but also opens up new possibilities in the field of AI-driven art and cultural interpretation, fostering a deeper connection with our past and inspiring a future where technology and culture converge to create meaningful experiences.

# Bibliography

[1] Radhia Abd Jelil, Xianyi Zeng, Ludovic Koehl, and Anne Perwuelz. Modeling plasma surface modification of textile fabrics using artificial neural networks. *Engineering Applications of Artificial Intelligence*, 26(8):1854–1864, 2013.

[2] Bishwo Adhikari and Heikki Huttunen. Iterative bounding box annotation for object detection. In *2020 25th International Conference on Pattern Recognition (ICPR)*, pages 4040–4046. IEEE, 2021.

[3] Kripesh Adhikari, Hamid Bouchachia, and Hammadi Nait-Charif. Activity recognition for indoor fall detection using convolutional neural network. In *2017 Fifteenth IAPR International Conference on Machine Vision Applications (MVA)*, pages 81–84. IEEE, 2017.

[4] Misbah Ahmad, Imran Ahmed, Kaleem Ullah, and Maaz Ahmad. A deep neural network approach for top view people detection and counting. In *2019 IEEE 10th Annual Ubiquitous Computing, Electronics & Mobile Communication Conference (UEMCON)*, pages 1082–1088. IEEE, 2019.

[5] Dror Aiger and Hugues Talbot. The phase only transform for unsupervised surface defect detection. In *2010 IEEE Computer Society Conference on Computer Vision and Pattern Recognition*, pages 295–302. IEEE, 2010.

[6] Mykhaylo Andriluka, Leonid Pishchulin, Peter Gehler, and Bernt Schiele. 2d human pose estimation: New benchmark and state of the art analysis. In *Proceedings of the IEEE Conference on Computer Vision and Pattern Recognition*, pages 3686–3693, 2014.

[7] Martin Arjovsky, Soumith Chintala, and Léon Bottou. Wasserstein generative adversarial networks. In *International Conference On Machine Learning*, pages 214–223. PMLR, 2017.

[8] M. Aslan, A. Sengur, Y. Xiao, H. Wang, M. C. Ince, and X. Ma. Shape feature encoding via fisher vector for efficient fall detection in depth-videos. *Appl. Soft Comput.*, 37:1023–1028, 2015.

[9] Haoyue Bai, Song Wen, and S-H Gary Chan. Crowd counting on images with scale variation and isolated clusters. In *2019 IEEE/CVF International Conference On Computer Vision Workshop (ICCVW)*, pages 18–27. IEEE, 2019.

[10] Liron Bergman, Niv Cohen, and Yedid Hoshen. Deep nearest neighbor anomaly detection. *ArXiv Preprint ArXiv:2002.10445*, 2020.

[11] Paul Bergmann, Michael Fauser, David Sattlegger, and Carsten Steger. Mvtec ad – a comprehensive real-world dataset for unsupervised anomaly detection. In *Proceedings of the IEEE/CVF Conference on Computer Vision and Pattern Recognition*, pages 9592–9600, 2019.

[12] Hakan Bilen, Basura Fernando, Efstratios Gavves, and Andrea Vedaldi. Action recognition with dynamic image networks. *IEEE Transactions on Pattern Analysis and Machine Intelligence*, 40(12):2799–2813, 2017.

[13] Alexey Bochkovskiy, Chien-Yao Wang, and Hong-Yuan Mark Liao. Yolov4: Optimal speed and accuracy of object detection. *ArXiv Preprint ArXiv: 2004.10934*, 2020.

[14] Avishek Joey Bose and Parham Aarabi. Adversarial attacks on face detectors using neural net based constrained optimization. In *2018 IEEE 20th International Workshop on Multimedia Signal Processing (MMSP)*, pages 1–6. IEEE, 2018.

[15] Andrew Brock. Large scale GAN training for high fidelity natural image synthesis. *ArXiv Preprint ArXiv:1809.11096*, 2018.

[16] Tom Brown, Benjamin Mann, Nick Ryder, Melanie Subbiah, Jared D Kaplan, Prafulla Dhariwal, Arvind Neelakantan, Pranav Shyam, Girish Sastry, Amanda Askell, et al. Language models are few-shot learners. *Advances in Neural Information Processing Systems*, 33:1877–1901, 2020.

[17] Michael Buckland and Fredric Gey. The relationship between recall and precision. *Journal of the American Society for Information Science*, 45(1):12–19, 1994.

[18] Yindi Cai, Qi Sang, Zhi-Feng Lou, and Kuang-Chao Fan. Error analysis and compensation of a laser measurement system for simultaneously measuring five-degree-of-freedom error motions of linear stages. *Sensors*, 19(18):3833, 2019.

[19] Xinkun Cao, Zhipeng Wang, Yanyun Zhao, and Fei Su. Scale aggregation network for accurate and efficient crowd counting. In *Proceedings of the European Conference on Computer Vision (ECCV)*, pages 734–750, 2018.

[20] Z. Cao, G. Hidalgo Martinez, T. Simon, S. Wei, and Y. A. Sheikh. Openpose: Realtime multi-person 2d pose estimation using part affinity fields. *IEEE Transactions On Pattern Analysis And Machine Intelligence*, 2019.

[21] Zhe Cao, Tomas Simon, Shih-En Wei, and Yaser Sheikh. Realtime multi-person 2d pose estimation using part affinity fields. In *Proceedings of the IEEE Conference On Computer Vision And Pattern Recognition*, pages 7291–7299, 2017.

[22] Antoni B Chan and Nuno Vasconcelos. Bayesian poisson regression for crowd counting. In *2009 IEEE 12th International Conference on Computer Vision*, pages 545–551. IEEE, 2009.

[23] Antoni B Chan and Nuno Vasconcelos. Counting people with low-level features and bayesian regression. *IEEE Transactions on Image Processing*, 21(4):2160–2177, 2011.

[24] Kai Chen, Jiaqi Wang, Jiangmiao Pang, Yuhang Cao, Yu Xiong, Xiaoxiao Li, Shuyang Sun, Wansen Feng, Ziwei Liu, Jiarui Xu, et al. Mmdetection: Open mmlab detection toolbox and benchmark. *ArXiv Preprint ArXiv:1906.07155*, 2019.

[25] Ke Chen, Shaogang Gong, Tao Xiang, and Chen Change Loy. Cumulative attribute space for age and crowd density estimation. In *Proceedings of the IEEE Conference on Computer Vision and Pattern Recognition*, pages 2467–2474, 2013.

[26] Ke Chen, Chen Change Loy, Shaogang Gong, and Tony Xiang. Feature mining for localised crowd counting. In *Bmvc*, volume 1, page 3, 2012.

[27] Lei Chen, Ming Yang, Luguo Hao, and Danda Rawat. Framework and challenges: H. 265/hevc rate control in real-time transmission over 5g mobile networks. In *10th EAI International Conference on Mobile Multimedia Communications*, pages 192–198, 2017.

[28] Liang-Chieh Chen, George Papandreou, Florian Schroff, and Hartwig Adam. Rethinking atrous convolution for semantic image segmentation. *ArXiv Preprint ArXiv:1706.05587*, 2017.

[29] Liang-Chieh Chen, Yukun Zhu, George Papandreou, Florian Schroff, and Hartwig Adam. Encoder-decoder with atrous separable convolution for semantic image segmentation. In *Proceedings of the European Conference On Computer Vision (ECCV)*, pages 801–818, 2018.

[30] Xi Chen, Yan Duan, Rein Houthooft, John Schulman, Ilya Sutskever, and Pieter Abbeel. Infogan: Interpretable representation learning by information maximizing generative adversarial nets. *Advances in Neural Information Processing Systems*, 29, 2016.

[31] Xinlei Chen, Hao Fang, Tsung-Yi Lin, Ramakrishna Vedantam, Saurabh Gupta, Piotr Dollár, and C Lawrence Zitnick. Microsoft coco captions: Data collection and evaluation server. *ArXiv Preprint ArXiv:1504.00325*, 2015.

[32] Daniell Chiang. Detect faces and determine whether people are wearing mask. *Face Mask Detection*, 15, 2020.

[33] Chin-Jou Chong, Wooi-Haw Tan, Yoong Choon Chang, Mohamed Farid Noor Batcha, and Ettikan Karuppiah. Visual based fall detection with reduced complexity horprasert segmentation using superpixel. In *2015 IEEE 12th International Conference on Networking, Sensing and Control*, pages 462–467. IEEE, 2015.

[34] Aakanksha Chowdhery, Sharan Narang, Jacob Devlin, Maarten Bosma, Gaurav Mishra, Adam Roberts, Paul Barham, Hyung Won Chung, Charles Sutton, Sebastian Gehrmann, et al. Palm: Scaling language modeling with pathways. *Journal of Machine Learning Research*, 24(240):1–113, 2023.

[35] Niv Cohen and Yedid Hoshen. Sub-image anomaly detection with deep pyramid correspondences. *ArXiv Preprint ArXiv:2005.02357*, 2020.

[36] Shuang Cong and Yang Zhou. A review of convolutional neural network architectures and their optimizations. *Artificial Intelligence Review*, 56(3):1905–1969, 2023.

[37] Corinna Cortes and Vladimir Vapnik. Support-vector networks. *Machine Learning*, 20(3):273–297, 1995.

[38] Florinel-Alin Croitoru, Vlad Hondru, Radu Tudor Ionescu, and Mubarak Shah. Diffusion models in vision: A survey. *IEEE Transactions on Pattern Analysis And Machine Intelligence*, 45(9):10850–10869, 2023.

[39] Yi Cui, Fuqiang Zhou, Yexin Wang, Liu Liu, and He Gao. Precise calibration of binocular vision system used for vision measurement. *Optics Express*, 22(8):9134–9149, 2014.

[40] Zihang Dai, Zhilin Yang, Yiming Yang, Jaime Carbonell, Quoc V Le, and Ruslan Salakhutdinov. Transformer-xl: Attentive language models beyond a fixed-length context. *ArXiv Preprint ArXiv:1901.02860*, 2019.

[41] Koldo De Miguel, Alberto Brunete, Miguel Hernando, and Ernesto Gambao. Home camera-based fall detection system for the elderly. *Sensors*, 17(12):2864, 2017.

[42] Jia Deng, Wei Dong, Richard Socher, Li-Jia Li, Kai Li, and Li Fei-Fei. Imagenet: A large-scale hierarchical image database. In *2009 IEEE Conference on Computer Vision and Pattern Recognition*, pages 248–255. IEEE, 2009.

[43] Jiankang Deng, Jia Guo, Niannan Xue, and Stefanos Zafeiriou. Arcface: Additive angular margin loss for deep face recognition. In *Proceedings of the IEEE/CVF Conference on Computer Vision and Pattern Recognition*, pages 4690–4699, 2019.

[44] Jacob Devlin, Ming-Wei Chang, Kenton Lee, and Kristina Toutanova. Bert: Pre-training of deep bidirectional transformers for language understanding. *ArXiv Preprint ArXiv:1810.04805*, 2018.

[45] Piotr Dollár, Zhuowen Tu, Pietro Perona, and Serge Belongie. *Integral Channel Features*, page 91.1–91.11. BMVC Press, 2009. Funding by NSF.

[46] Jeffrey Donahue, Lisa Anne Hendricks, Sergio Guadarrama, Marcus Rohrbach, Subhashini Venugopalan, Kate Saenko, and Trevor Darrell. Long-term recurrent

convolutional networks for visual recognition and description. In *Proceedings of the IEEE Conference on Computer Vision and Pattern Recognition*, pages 2625–2634, 2015.

[47] Shi Dong, Ping Wang, and Khushnood Abbas. A survey on deep learning and its applications. *Computer Science Review*, 40:100379, 2021.

[48] Alexey Dosovitskiy, Lucas Beyer, Alexander Kolesnikov, Dirk Weissenborn, Xiaohua Zhai, Thomas Unterthiner, Mostafa Dehghani, Matthias Minderer, Georg Heigold, Sylvain Gelly, et al. An image is worth 16x16 words: Transformers for image recognition at scale. *ArXiv Preprint ArXiv:2010.11929*, 2020.

[49] Matthijs Douze, Alexandr Guzhva, Chengqi Deng, Jeff Johnson, Gergely Szilvasy, Pierre-Emmanuel Mazar'e, Maria Lomeli, Lucas Hosseini, and Herv'e J'egou. The faiss library. *ArXiv*, abs/2401.08281, 2024.

[50] Wenbin Du, Yali Wang, and Yu Qiao. Rpan: An end-to-end recurrent pose-attention network for action recognition in videos. In *Proceedings of the IEEE International Conference on Computer Vision*, pages 3725–3734, 2017.

[51] Kaiwen Duan, Song Bai, Lingxi Xie, Honggang Qi, Qingming Huang, and Qi Tian. Centernet: Keypoint triplets for object detection. In *Proceedings of the IEEE/CVF International Conference on Computer Vision*, pages 6569–6578, 2019.

[52] M. Everingham, S. M. Eslami, L. Gool, C. K. Williams, J. Winn, and A. Zisserman. The pascal visual object classes challenge: A retrospective. *International Journal of Computer Vision*, 2015.

[53] Mark Everingham, SM Eslami, Luc Van Gool, Christopher KI Williams, John Winn, and Andrew Zisserman. The pascal visual object classes challenge: A retrospective. *International Journal of Computer Vision*, 111(1):98–136, 2015.

[54] Mark Everingham, Luc Van Gool, Christopher KI Williams, John Winn, and Andrew Zisserman. The pascal visual object classes (voc) challenge. *International Journal of Computer Vision*, 88:303–338, 2010.

[55] Kevin Eykholt, Ivan Evtimov, Earlence Fernandes, Bo Li, Amir Rahmati, Chaowei Xiao, Atul Prakash, Tadayoshi Kohno, and Dawn Song. Robust physical-world attacks on deep learning visual classification. In *Proceedings of the IEEE Conference on Computer Vision and Pattern Recognition*, pages 1625–1634, 2018.

[56] Y. Fan, M. D. Levine, G. Wen, and S. Qiu. A deep neural network for real-time detection of falling humans in naturally occurring scenes. *Neurocomputing*, 260:43–58, 2017.

[57] Lunke Fei, Guangming Lu, Wei Jia, Shaohua Teng, and David Zhang. Feature extraction methods for palmprint recognition: A survey and evaluation. *IEEE Transactions on Systems, Man, and Cybernetics: Systems*, 49(2):346–363, 2018.

[58] Pedro F Felzenszwalb, Ross B Girshick, David McAllester, and Deva Ramanan. Object detection with discriminatively trained part-based models. *IEEE Transactions on Pattern Analysis and Machine Intelligence*, 32(9):1627–1645, 2010.

[59] JW Funck, Y Zhong, DA Butler, CC Brunner, and JB Forrer. Image segmentation algorithms applied to wood defect detection. *Computers and Electronics in Agriculture*, 41(1-3):157–179, 2003.

[60] Shiming Ge, Jia Li, Qiting Ye, and Zhao Luo. Detecting masked faces in the wild with lle-cnns. In *2017 IEEE Conference on Computer Vision and Pattern Recognition (CVPR)*, 2017.

[61] Weina Ge and Robert T Collins. Marked point processes for crowd counting. In *2009 IEEE Conference on Computer Vision and Pattern Recognition*, pages 2913–2920. IEEE, 2009.

[62] Zheng Ge, Songtao Liu, Feng Wang, Zeming Li, and Jian Sun. Yolox: Exceeding yolo series in 2021. *ArXiv Preprint ArXiv:2107.08430*, 2021.

[63] Ross Girshick. Fast r-cnn. In *Proceedings of the IEEE International Conference on Computer Vision*, pages 1440–1448, 2015.

[64] Ross Girshick, Jeff Donahue, Trevor Darrell, and Jitendra Malik. Rich feature hierarchies for accurate object detection and semantic segmentation. In *Proceedings of the IEEE Conference on Computer Vision and Pattern Recognition*, pages 580–587, 2014.

[65] Ross Girshick, Jeff Donahue, Trevor Darrell, and Jitendra Malik. Rich feature hierarchies for accurate object detection and semantic segmentation. In *Proceedings of the IEEE Conference on Computer Vision and Pattern Recognition*, pages 580–587, 2014.

[66] Simon Gloser, Marcel Soulier, and Luis A Tercero Espinoza. Dynamic analysis of global copper flows. global stocks, postconsumer material flows, recycling indicators, and uncertainty evaluation. *Environmental Science & Technology*, 47(12):6564–6572, 2013.

[67] Fernando Gomez, Juan Ignacio Guzmán, and John E Tilton. Copper recycling and scrap availability. *Resources Policy*, 32(4):183–190, 2007.

[68] Jia Gong, Zhipeng Fan, Qiuhong Ke, Hossein Rahmani, and Jun Liu. Meta agent teaming active learning for pose estimation. In *Proceedings of the IEEE/CVF Conference on Computer Vision and Pattern Recognition*, pages 11079–11089, 2022.

[69] Ian Goodfellow, Yoshua Bengio, and Aaron Courville. *Deep Learning*. MIT press, 2016.

[70] Ian Goodfellow, Jean Pouget-Abadie, Mehdi Mirza, Bing Xu, David Warde-Farley, Sherjil Ozair, Aaron Courville, and Yoshua Bengio. Generative adversarial nets. *Advances in Neural Information Processing Systems*, 27, 2014.

[71] Ian Goodfellow, Jean Pouget-Abadie, Mehdi Mirza, Bing Xu, David Warde-Farley, Sherjil Ozair, Aaron Courville, and Yoshua Bengio. Generative adversarial networks. *Communications of the ACM*, 63(11):139–144, 2020.

[72] Ian J Goodfellow, Jonathon Shlens, and Christian Szegedy. Explaining and harnessing adversarial examples. *ArXiv Preprint ArXiv:1412.6572*, 2014.

[73] Kishnaprasad G Gunale and Prachi Mukherji. Fall detection using k-nearest neighbor classification for patient monitoring. In *2015 International Conference on Information Processing (ICIP)*, pages 520–524. IEEE, 2015.

[74] Kai Han, Yunhe Wang, Hanting Chen, Xinghao Chen, Jianyuan Guo, Zhenhua Liu, Yehui Tang, An Xiao, Chunjing Xu, Yixing Xu, et al. A survey on vision transformer. *IEEE Transactions on Pattern Analysis and Machine Intelligence*, 45(1):87–110, 2022.

[75] Xian-Feng Han, Hamid Laga, and Mohammed Bennamoun. Image-based 3D Object Reconstruction: State-of-the-Art and Trends in the Deep Learning Era. *arXiv e-prints*, page arXiv:1906.06543, June 2019.

[76] Kazım Hanbay, Muhammed Fatih Talu, and Ömer Faruk Özgüven. Fabric defect detection systems and methods—a systematic literature review. *Optik*, 127(24):11960–11973, 2016.

[77] Kaiming He, Xiangyu Zhang, Shaoqing Ren, and Jian Sun. Spatial pyramid pooling in deep convolutional networks for visual recognition. *IEEE Transactions on Pattern Analysis and Machine Intelligence (TPAMI)*, 37(9):1904–1916, 2015. 2, 4, 7.

[78] Kaiming He, Xiangyu Zhang, Shaoqing Ren, and Jian Sun. Deep residual learning for image recognition. In *Proceedings of the IEEE Conference on Computer Vision and Pattern Recognition*, pages 770–778, 2016.

[79] Paul Henderson and Vittorio Ferrari. End-to-end training of object class detectors for mean average precision. In *Computer Vision–ACCV 2016: 13th Asian Conference on Computer Vision, Taipei, Taiwan, November 20-24, 2016, Revised Selected Papers, Part V 13*, pages 198–213. Springer, 2017.

[80] Geoffrey E. Hinton, Simon Osindero, and Yee Whye Teh. A fast learning algorithm for deep belief nets. *Neural Computation*, 18(7):1527–1554, 2006.

[81] Jonathan Ho, Ajay Jain, and Pieter Abbeel. Denoising diffusion probabilistic models. *Advances in Neural Information Processing Systems*, 33:6840–6851, 2020.

[82] Guang-Hua Hu, Qing-Hui Wang, and Guo-Hui Zhang. Unsupervised defect detection in textiles based on fourier analysis and wavelet shrinkage. *Applied Optics*, 54(10):2963–2980, 2015.

[83] Chenn-Jung Huang. Clustered defect detection of high quality chips using self-supervised multilayer perceptron. *Expert Systems with Applications*, 33(4):996–1003, 2007.

[84] Lianghua Huang, Wei Wang, Zhi-Fan Wu, Yupeng Shi, Huanzhang Dou, Chen Liang, Yutong Feng, Yu Liu, and Jingren Zhou. In-context lora for diffusion transformers. *ArXiv Preprint ArXiv:2410.23775*, 2024.

[85] Rachel Huang, Jonathan Pedoeem, and Cuixian Chen. Yolo-lite: a real-time object detection algorithm optimized for non-gpu computers. In *2018 IEEE International Conference on Big Data (Big Data)*, pages 2503–2510. IEEE, 2018.

[86] Zhangjin Huang, Yuxin Wen, Zihao Wang, Jinjuan Ren, and Kui Jia. Surface reconstruction from point clouds: A survey and a benchmark. *ArXiv Preprint ArXiv:2205.02413*, 2022.

[87] Muhammad Hussain. Yolov1 to v8: Unveiling each variant–a comprehensive review of yolo. *IEEE Access*, 12:42816–42833, 2024.

[88] Haroon Idrees, Imran Saleemi, Cody Seibert, and Mubarak Shah. Multi-source multi-scale counting in extremely dense crowd images. In *Proceedings of the IEEE Conference on Computer Vision and Pattern Recognition*, pages 2547–2554, 2013.

[89] Haroon Idrees, Muhmmad Tayyab, Kishan Athrey, Dong Zhang, Somaya Al-Maadeed, Nasir Rajpoot, and Mubarak Shah. Composition loss for counting, density map estimation and localization in dense crowds. In *Proceedings of the European Conference on Computer Vision (ECCV)*, pages 532–546, 2018.

[90] Mohammadreza Iman, Hamid Reza Arabnia, and Khaled Rasheed. A review of deep transfer learning and recent advancements. *Technologies*, 11(2):40, 2023.

[91] Umar Iqbal, Pavlo Molchanov, Thomas Breuel Juergen Gall, and Jan Kautz. Hand pose estimation via latent 2.5 d heatmap regression. In *Proceedings of the European Conference on Computer Vision (ECCV)*, pages 118–134, 2018.

[92] Shahram Izadi, David Kim, Otmar Hilliges, David Molyneaux, Richard Newcombe, Pushmeet Kohli, Jamie Shotton, Steve Hodges, Dustin Freeman, Andrew Davison, et al. Kinectfusion: real-time 3d reconstruction and interaction using a moving depth camera. In *Proceedings of the 24th Annual ACM Symposium on User Interface Software And Technology*, pages 559–568, 2011.

[93] Christian Janiesch, Patrick Zschech, and Kai Heinrich. Machine learning and deep learning. *Electronic Markets*, 31(3):685–695, 2021.

[94] Shuiwang Ji, Wei Xu, Ming Yang, and Kai Yu. 3d convolutional neural networks for human action recognition. *IEEE Transactions on Pattern Analysis and Machine Intelligence*, 35(1):221–231, 2012.

[95] Juyong Jiang, Fan Wang, Jiasi Shen, Sungju Kim, and Sunghun Kim. A survey on large language models for code generation. *ArXiv*, abs/2406.00515, 2024.

[96] Peiyuan Jiang, Daji Ergu, Fangyao Liu, Ying Cai, and Bo Ma. A review of yolo algorithm developments. *Procedia Computer Science*, 199:1066–1073, 2022.

[97] Pengfei Jing, Qiyi Tang, Yuefeng Du, Lei Xue, Xiapu Luo, Ting Wang, Sen Nie, and Shi Wu. Too good to be safe: Tricking lane detection in autonomous driving with crafted perturbations. In *30th USENIX Security Symposium (USENIX Security 21)*, pages 3237–3254, 2021.

[98] Glenn Jocher. Yolov5, 2020.

[99] Glenn Jocher, Ayush Chaurasia, and Jing Qiu. *Ultralytics YOLO*, January 2023.

[100] Glenn Jocher, Ayush Chaurasia, Alex Stoken, Jirka Borovec, Yonghye Kwon, Kalen Michael, Jiacong Fang, Zeng Yifu, Colin Wong, Diego Montes, et al. Ultralytics/yolov5: V7.0-yolov5 sota realtime instance segmentation. *Zenodo*, 2022.

[101] Glenn Jocher, Ayush Chaurasia, Alex Stoken, Jirka Borovec, NanoCode012, Yonghye Kwon, TaoXie, Jiacong Fang, imyhxy, Kalen Michael, Lorna, Abhiram V, Diego Montes, Jebastin Nadar, Laughing, tkianai, yxNONG, Piotr Skalski, Zhiqiang Wang, Adam Hogan, Cristi Fati, Lorenzo Mammana, AlexWang1900, Deep Patel, Ding Yiwei, Felix You, Jan Hajek, Laurentiu Diaconu, and Mai Thanh Minh. ultralytics/yolov5: v6.1—TensorRT, TensorFlow Edge TPU and OpenVINO Export and Inference, February 2022.

[102] L. H. Juang and M. N. Wu. Fall down detection under smart home system. *J. Med. Syst.*, 39:107–113, 2015.

[103] Zhiqiang Kang, Chaohui Yuan, and Qian Yang. The fabric defect detection technology based on wavelet transform and neural network convergence. In *2013 IEEE International Conference on Information and Automation (ICIA)*, pages 597–601. IEEE, 2013.

[104] Andrej Karpathy, George Toderici, Sanketh Shetty, Thomas Leung, Rahul Sukthankar, and Li Fei-Fei. Large-scale video classification with convolutional neural networks. In *Proceedings of the IEEE Conference on Computer Vision and Pattern Recognition*, pages 1725–1732, 2014.

[105] Tero Karras. A style-based generator architecture for generative adversarial networks. *ArXiv Preprint ArXiv:1812.04948*, 2019.

[106] Tero Karras, Samuli Laine, Miika Aittala, Janne Hellsten, Jaakko Lehtinen, and Timo Aila. Analyzing and improving the image quality of stylegan. In *Proceedings of the IEEE/CVF Conference on Computer Vision and Pattern Recognition*, pages 8110–8119, 2020.

[107] Salman Khan, Muzammal Naseer, Munawar Hayat, Syed Waqas Zamir, Fahad Shahbaz Khan, and Mubarak Shah. Transformers in vision: A survey. *ACM Computing Surveys (CSUR)*, 54(10s):1–41, 2022.

[108] Bongjun Kim and Bryan Pardo. A human-in-the-loop system for sound event detection and annotation. *ACM Transactions on Interactive Intelligent Systems (TiiS)*, 8(2):1–23, 2018.

[109] D. P. Kingma and J. Ba. Adam: A method for stochastic optimization. *ArXiv Preprint ArXiv:1412.6980*, 2014.

[110] Diederik P. Kingma and Jimmy Ba. Adam: A method for stochastic optimization. *ArXiv E-prints*, 2014.

[111] Alexander Kirillov, Eric Mintun, Nikhila Ravi, Hanzi Mao, Chloe Rolland, Laura Gustafson, Tete Xiao, Spencer Whitehead, Alexander C Berg, Wan-Yen Lo, et al. Segment anything. In *Proceedings of the IEEE/CVF International Conference on Computer Vision*, pages 4015–4026, 2023.

[112] Alex Krizhevsky, I. Sutskever, and G. Hinton. Imagenet classification with deep convolutional neural networks. *Advances in Neural Information Processing Systems*, 25(2), 2012.

[113] Ajay Kumar. Computer-vision-based fabric defect detection: A survey. *IEEE Transactions on Industrial Electronics*, 55(1):348–363, 2008.

[114] Bogdan Kwolek and Michal Kepski. Human fall detection on embedded platform using depth maps and wireless accelerometer. *Computer Methods and Programs in Biomedicine*, 117(3):489–501, December 2014. ISSN 0169-2607.

[115] Hei Law and Jia Deng. Cornernet: Detecting objects as paired keypoints. In *Proceedings of the European Conference on Computer Vision (ECCV)*, pages 734–750, 2018.

[116] Hei Law, Yun Teng, Olga Russakovsky, and Jia Deng. Cornernet-lite: Efficient keypoint based object detection. arXiv preprint 2, 2019.

[117] Yann LeCun, Yoshua Bengio, and Geoffrey Hinton. Deep learning. *Nature*, 521(7553):436–444, 2015.

[118] Yann LeCun, Bernhard Boser, John S Denker, Donnie Henderson, Richard E Howard, Wayne Hubbard, and Lawrence D Jackel. Backpropagation applied to handwritten zip code recognition. *Neural Computation*, 1(4):541–551, 1989.

[119] Yann LeCun, Léon Bottou, Yoshua Bengio, and Patrick Haffner. Gradient-based learning applied to document recognition. *Proceedings of the IEEE*, 86(11):2278–2324, 1998.

[120] Kok Tong Lee, Pei Song Chee, Eng Hock Lim, and Chu Chen Lim. Artificial intelligence (ai)-driven smart glove for object recognition application. *Materials Today: Proceedings*, 64:1563–1568, 2022.

[121] Victor Lempitsky and Andrew Zisserman. Learning to count objects in images. *Advances in Neural Information Processing Systems*, 23, 2010.

[122] Chun-Liang Li, Kihyuk Sohn, Jinsung Yoon, and Tomas Pfister. Cutpaste: Self-supervised learning for anomaly detection and localization. In *Proceedings of the IEEE/CVF Conference on Computer Vision and Pattern Recognition*, pages 9664–9674, 2021.

[123] Chuyi Li, Lulu Li, Hongliang Jiang, Kaiheng Weng, Yifei Geng, Liang Li, Zaidan Ke, Qingyuan Li, Meng Cheng, Weiqiang Nie, et al. Yolov6: A single-stage object detection framework for industrial applications. *ArXiv Preprint ArXiv:2209.02976*, 2022.

[124] Hao Li, Jiaohua Qin, Xuyu Xiang, Lili Pan, Wentao Ma, and Neal N Xiong. An efficient image matching algorithm based on adaptive threshold and ransac. *IEEE Access*, 6:66963–66971, 2018.

[125] Hong-an Li, Liuqing Hu, Jun Liu, Jing Zhang, and Tian Ma. A review of advances in image inpainting research. *The Imaging Science Journal*, 72(5):669–691, 2024.

[126] Min Li, Zhaoxiang Zhang, Kaiqi Huang, and Tieniu Tan. Estimating the number of people in crowded scenes by mid based foreground segmentation and head-shoulder detection. In *2008 19th International Conference on Pattern Recognition*, pages 1–4. IEEE, 2008.

[127] Xiaogang Li, Tiantian Pang, Weixiang Liu, and Tianfu Wang. Fall detection for elderly person care using convolutional neural networks. In *2017 10th International Congress On Image And Signal Processing, Biomedical Engineering And Informatics (CISP-BMEI)*, pages 1–6. IEEE, 2017.

[128] Yuhong Li, Xiaofan Zhang, and Deming Chen. Csrnet: Dilated convolutional neural networks for understanding the highly congested scenes. In *Proceedings of the IEEE Conference on Computer Vision and Pattern Recognition*, pages 1091–1100, 2018.

[129] Yundong Li, Weigang Zhao, and Jiahao Pan. Deformable patterned fabric defect detection with fisher criterion-based deep learning. *IEEE Transactions on Automation Science and Engineering*, 14(2):1256–1264, 2016.

[130] Zewen Li, Fan Liu, Wenjie Yang, Shouheng Peng, and Jun Zhou. A survey of convolutional neural networks: analysis, applications, and prospects. *IEEE Transactions on Neural Networks and Learning Systems*, 33(12):6999–7019, 2021.

[131] Sheng-Fuu Lin, Jaw-Yeh Chen, and Hung-Xin Chao. Estimation of number of people in crowded scenes using perspective transformation. *IEEE Transactions on Systems, Man, and Cybernetics-Part A: Systems and Humans*, 31(6):645–654, 2001.

[132] T. Y. Lin et al. Microsoft coco: Common objects in context. In *European Conference on Computer Vision*, pages 740–755, 2014.

[133] Tsung Yi Lin, Michael Maire, Serge Belongie, James Hays, and C. Lawrence Zitnick. Microsoft coco: Common objects in context. *Springer International Publishing*, 2014.

[134] Weiyao Lin, Ming-Ting Sun, Radha Poovandran, and Zhengyou Zhang. Human activity recognition for video surveillance. In *2008 IEEE International Symposium on Circuits and Systems (ISCAS)*, pages 2737–2740. IEEE, 2008.

[135] Zhe Lin and Larry S Davis. Shape-based human detection and segmentation via hierarchical part-template matching. *IEEE Transactions on Pattern Analysis and Machine Intelligence*, 32(4):604–618, 2010.

[136] Jian Liu, Hossein Rahmani, Naveed Akhtar, and Ajmal Mian. Learning human pose models from synthesized data for robust RGB-D action recognition. *International Journal of Computer Vision*, 127:1545–1564, 2019.

[137] Jun Liu, Amir Shahroudy, Dong Xu, Alex C Kot, and Gang Wang. Skeleton-based action recognition using spatio-temporal LSTM network with trust gates. *IEEE Transactions on Pattern Analysis and Machine Intelligence*, 40(12):3007–3021, 2017.

[138] Shu Liu, Lu Qi, Haifang Qin, Jianping Shi, and Jiaya Jia. Path aggregation network for instance segmentation. In *Proceedings of the IEEE Conference on Computer Vision and Pattern Recognition*, pages 8759–8768, 2018.

[139] Wei Liu, Dragomir Anguelov, Dumitru Erhan, Christian Szegedy, Scott Reed, Cheng-Yang Fu, and Alexander C Berg. Ssd: Single shot multibox detector. In *Computer Vision–ECCV 2016: 14th European Conference, Amsterdam, The Netherlands, October 11–14, 2016, Proceedings, Part I 14*, pages 21–37. Springer, 2016.

[140] Yinhan Liu, Myle Ott, Naman Goyal, Jingfei Du, Mandar Joshi, Danqi Chen, Omer Levy, Mike Lewis, Luke Zettlemoyer, and Veselin Stoyanov. Roberta: A robustly optimized bert pretraining approach. *ArXiv*, abs/1907.11692, 2019.

[141] Yixin Liu, Kai Zhang, Yuan Li, Zhiling Yan, Chujie Gao, Ruoxi Chen, Zhengqing Yuan, Yue Huang, Hanchi Sun, Jianfeng Gao, Lifang He, and Lichao Sun. Sora: A review on background, technology, limitations, and opportunities of large vision models. *ArXiv*, abs/2402.17177, 2024.

[142] Ze Liu, Yutong Lin, Yue Cao, Han Hu, Yixuan Wei, Zheng Zhang, Stephen Lin, and Baining Guo. Swin transformer: Hierarchical vision transformer using shifted windows. In *Proceedings Of The IEEE/CVF International Conference On Computer Vision*, pages 10012–10022, 2021.

[143] Stuart Lloyd. Least squares quantization in pcm. *IEEE Transactions on Information Theory*, 28(2):129–137, 1982.

[144] Jonathan Long, Evan Shelhamer, and Trevor Darrell. Fully convolutional networks for semantic segmentation. In *Proceedings of the IEEE Conference On Computer Vision and Pattern Recognition*, pages 3431–3440, 2015.

[145] Xiang Long, Kaipeng Deng, Guanzhong Wang, Yang Zhang, Qingqing Dang, Yuan Gao, Hui Shen, Jianguo Ren, Shumin Han, Errui Ding, et al. PP-YOLO: An effective and efficient implementation of object detector. arXiv preprint 2, 7, 2020.

[146] Mingqi Lu, Yaocong Hu, and Xiaobo Lu. Driver action recognition using deformable and dilated faster r-cnn with optimized region proposals. *Applied Intelligence*, 50(4):1100–1111, 2020.

[147] Zhiheng Ma, Xing Wei, Xiaopeng Hong, and Yihong Gong. Bayesian loss for crowd count estimation with point supervision. In *Proceedings of the IEEE/CVF International Conference on Computer Vision*, pages 6142–6151, 2019.

[148] Abdelhak Mahmoudi and Fakhita Regragui. Welding defect detection by segmentation of radiographic images. In *2009 WRI World Congress on Computer Science and Information Engineering*, volume 7, pages 111–115. IEEE, 2009.

[149] Amitha Mathew, P Amudha, and S Sivakumari. Deep learning techniques: an overview. *Advanced Machine Learning Technologies and Applications: Proceedings of AMLTA 2020*, pages 599–608, 2021.

[150] Warren S. Mcculloch and Walter Pitts. A logical calculus of the ideas immanent in nervous activity. *Biol Math Biophys*, pages 115–133, 1943.

[151] Dushyant Mehta, Andrii Skliar, Haitam Ben Yahia, Shubhankar Borse, Fatih Porikli, Amirhossein Habibian, and Tijmen Blankevoort. Simple and efficient architectures for semantic segmentation. In *Proceedings of the IEEE/CVF Conference on Computer Vision and Pattern Recognition*, pages 2628–2636, 2022.

[152] Anton Milan, Laura Leal-Taixé, Ian Reid, Stefan Roth, and Konrad Schindler. Mot16: A benchmark for multi-object tracking. *ArXiv Preprint ArXiv:1603.00831*, 2016.

[153] Minsky, Marvin, Papert, and Seymour. Perceptrons : An introduction to computational geometry. *The MIT Press*, 1969.

[154] Andrew Ng. Unbiggen ai. *IEEE Spectrum*, 9, 2022.

[155] Alexander Quinn Nichol and Prafulla Dhariwal. Improved denoising diffusion probabilistic models. In *International Conference on Machine Learning*, pages 8162–8171. PMLR, 2021.

[156] Shuteng Niu, Yongxin Liu, Jian Wang, and Houbing Song. A decade survey of transfer learning (2010–2020). *IEEE Transactions on Artificial Intelligence*, 1(2):151–166, 2020.

[157] Mark Nixon and Alberto Aguado. *Feature Extraction and Image Processing for Computer Vision*. Academic press, 2019.

[158] Ozan Oktay, Jo Schlemper, Loic Le Folgoc, Matthew Lee, Mattias Heinrich, Kazunari Misawa, Kensaku Mori, Steven McDonagh, Nils Y Hammerla, Bernhard Kainz, et al. Attention U-Net: Learning where to look for the pancreas. *ArXiv Preprint ArXiv:1804.03999*, 2018.

[159] OpenAI. ChatGPT: Optimizing language models for dialogue, 2022.

[160] Alessandro Ortis, Pasquale Caponnetto, Riccardo Polosa, Salvatore Urso, and Sebastiano Battiato. A report on smoking detection and quitting technologies. *International Journal of Environmental Research and Public Health*, 17(7):2614, 2020.

[161] Wanli Ouyang and Xiaogang Wang. Single-pedestrian detection aided by multi-pedestrian detection. In *Proceedings of the IEEE Conference on Computer Vision and Pattern Recognition*, pages 3198–3205, 2013.

[162] Sinno Jialin Pan and Qiang Yang. A survey on transfer learning. *IEEE Transactions on Knowledge and Data Engineering*, 22(10):1345–1359, 2010.

[163] Xingang Pan, Ayush Tewari, Thomas Leimkühler, Lingjie Liu, Abhimitra Meka, and Christian Theobalt. Drag your GAN: Interactive point-based manipulation on the generative image manifold. In *ACM SIGGRAPH 2023 Conference Proceedings*, pages 1–11, 2023.

[164] George Papandreou, Tyler Zhu, Nori Kanazawa, Alexander Toshev, Jonathan Tompson, Chris Bregler, and Kevin Murphy. Towards accurate multi-person pose estimation in the wild. In *Proceedings of the IEEE Conference on Computer Vision and Pattern Recognition*, pages 4903–4911, 2017.

[165] Adam Paszke, Sam Gross, Francisco Massa, Adam Lerer, James Bradbury, Gregory Chanan, Trevor Killeen, Zeming Lin, Natalia Gimelshein, Luca Antiga, et al. Pytorch: An imperative style, high-performance deep learning library. *Advances in Neural Information Processing Systems*, 32, 2019.

[166] Genevieve Patterson and James Hays. Coco attributes: Attributes for people, animals, and objects. In *Computer Vision–ECCV 2016: 14th European Conference, Amsterdam, The Netherlands, October 11-14, 2016, Proceedings, Part VI 14*, pages 85–100. Springer, 2016.

[167] Viet-Quoc Pham, Tatsuo Kozakaya, Osamu Yamaguchi, and Ryuzo Okada. Count forest: Co-voting uncertain number of targets using random forest for crowd density estimation. In *Proceedings of the IEEE International Conference on Computer Vision*, pages 3253–3261, 2015.

[168] Lucas Pinheiro Cinelli, Matheus Araújo Marins, Eduardo Antúnio Barros da Silva, and Sérgio Lima Netto. Variational autoencoder. In *Variational Methods for Machine Learning with Applications to Deep Networks*, pages 111–149. Springer, 2021.

[169] Dustin Podell, Zion English, Kyle Lacey, A. Blattmann, Tim Dockhorn, Jonas Muller, Joe Penna, and Robin Rombach. SDXL: Improving latent diffusion models for high-resolution image synthesis. *ArXiv*, abs/2307.01952, 2023.

[170] Ronald Poppe. A survey on vision-based human action recognition. *Image and Vision Computing*, 28(6):976–990, 2010.

[171] Ivan Prodaiko. Person re-identification in a top-view multi-camera environment. 2020.

[172] Vincent Rabaud and Serge Belongie. Counting crowded moving objects. In *2006 IEEE Computer Society Conference on Computer Vision and Pattern Recognition (CVPR'06)*, volume 1, pages 705–711. IEEE, 2006.

[173] Alec Radford. Unsupervised representation learning with deep convolutional generative adversarial networks. *ArXiv Preprint ArXiv:1511.06434*, 2015.

[174] Alec Radford, Karthik Narasimhan, Tim Salimans, Ilya Sutskever, et al. Improving language understanding by generative pre-training. 2018.

[175] Alec Radford, Jeffrey Wu, Rewon Child, David Luan, Dario Amodei, Ilya Sutskever, et al. Language models are unsupervised multitask learners. *OpenAI Blog*, 1(8):9, 2019.

[176] Colin Raffel, Noam Shazeer, Adam Roberts, Katherine Lee, Sharan Narang, Michael Matena, Yanqi Zhou, Wei Li, and Peter J Liu. Exploring the limits of transfer learning with a unified text-to-text transformer. *Journal of Machine Learning Research*, 21(140):1–67, 2020.

[177] Hossein Rahmani and Ajmal Mian. 3d action recognition from novel viewpoints. In *Proceedings of the IEEE Conference on Computer Vision and Pattern Recognition*, pages 1506–1515, 2016.

[178] Maryam Rahnemoonfar and Hend Alkittawi. Spatio-temporal convolutional neural network for elderly fall detection in depth video cameras. In *2018 IEEE International Conference on Big Data (Big Data)*, pages 2868–2873. IEEE, 2018.

[179] Joseph Redmon, Santosh Divvala, Ross Girshick, and Ali Farhadi. You only look once: Unified, real-time object detection. In *Proceedings of the IEEE Conference on Computer Vision and Pattern Recognition*, pages 779–788, 2016.

[180] Joseph Redmon and Ali Farhadi. Yolo9000: better, faster, stronger. In *Proceedings of the IEEE Conference on Computer Vision and Pattern Recognition*, pages 7263–7271, 2017.

[181] Joseph Redmon and Ali Farhadi. Yolov3: An incremental improvement. *ArXiv Preprint ArXiv:1804.02767*, 2018.

[182] Shaoqing Ren, Kaiming He, Ross Girshick, and Jian Sun. Faster R-CNN: Towards real-time object detection with region proposal networks. *Advances in Neural Information Processing Systems*, 28, 2015.

[183] Xiaofeng Ren. Multi-scale improves boundary detection in natural images. In *Computer Vision–ECCV 2008: 10th European Conference On Computer Vision, Marseille, France, October 12–18, 2008, Proceedings, Part III 10*, pages 533–545. Springer, 2008.

[184] Ergys Ristani, Francesco Solera, Roger Zou, Rita Cucchiara, and Carlo Tomasi. Performance measures and a data set for multi-target, multi-camera tracking. In *European Conference on Computer Vision*, pages 17–35. Springer, 2016.

[185] Daniel A Roberts, Sho Yaida, and Boris Hanin. *The principles of deep learning theory*. Cambridge University Press Cambridge, MA, USA, 2022.

[186] Robin Rombach, Andreas Blattmann, Dominik Lorenz, Patrick Esser, and Björn Ommer. High-resolution image synthesis with latent diffusion models. In *Proceedings of the IEEE/CVF Conference on Computer Vision and Pattern Recognition*, pages 10684–10695, 2022.

[187] Olaf Ronneberger, Philipp Fischer, and Thomas Brox. U-Net: Convolutional networks for biomedical image segmentation. In *Medical Image Computing and Computer-Assisted Intervention–MICCAI 2015: 18th International Conference, Munich, Germany, October 5–9, 2015, Proceedings, Part III 18*, pages 234–241. Springer, 2015.

[188] Rosenbilatt and F. The perceptron: a probabilistic model for information storage and organization in the brain. *Psychological Review*, 65:386–408, 1958.

[189] T-YLPG Ross and GKHP Dollár. Focal loss for dense object detection. In *Proceedings of the IEEE Conference on Computer Vision and Pattern Recognition*, pages 2980–2988, 2017.

[190] Karsten Roth, Latha Pemula, Joaquin Zepeda, Bernhard Schölkopf, Thomas Brox, and Peter Gehler. Towards total recall in industrial anomaly detection. In *Proceedings of the IEEE/CVF Conference on Computer Vision and Pattern Recognition*, pages 14318–14328, 2022.

[191] Biparnak Roy, Subhadip Nandy, Debojit Ghosh, Debarghya Dutta, Pritam Biswas, and Tamodip Das. Moxa: A deep learning based unmanned approach for real-time monitoring of people wearing medical masks. *Transactions of the Indian National Academy of Engineering*, 5:509–518, 2020.

[192] David E. Rumelhart, Geoffrey E. Hinton, and Ronald J. Williams. Learning representations by back propagating errors. *Nature*, 323(6088):533–536, 1986.

[193] Olga Russakovsky, Jia Deng, Hao Su, Jonathan Krause, Sanjeev Satheesh, Sean Ma, Zhiheng Huang, Andrej Karpathy, Aditya Khosla, Michael Bernstein, et al. Imagenet large scale visual recognition challenge. *International Journal of Computer Vision*, 115:211–252, 2015.

[194] Jürgen Schmidhuber. Deep learning in neural networks. *Neural Netw*, 2015.

[195] Johannes L Schonberger and Jan-Michael Frahm. Structure-from-motion revisited. In *Proceedings of the IEEE Conference on Computer Vision and Pattern Recognition*, pages 4104–4113, 2016.

[196] David M Scott. *Industrial process sensors*. CRC Press, 2018.

[197] Abdulkadir Şeker, Kadir Aşkın Peker, Ahmet Gürkan Yüksek, and Emre Delibaş. Fabric defect detection using deep learning. In *2016 24th Signal Processing and Communication Application Conference (SIU)*, pages 1437–1440. IEEE, 2016.

[198] Arindam Sengupta, Feng Jin, Renyuan Zhang, and Siyang Cao. mm-pose: Real-time human skeletal posture estimation using mmwave radars and cnns. *IEEE Sensors Journal*, 20(17):10032–10044, 2020.

[199] Volkan Y Senyurek, Masudul H Imtiaz, Prajakta Belsare, Stephen Tiffany, and Edward Sazonov. A CNN-LSTM neural network for recognition of puffing in smoking episodes using wearable sensors. *Biomedical Engineering Letters*, 10(2):195–203, 2020.

[200] Vaishaal Shankar, Rebecca Roelofs, Horia Mania, Alex Fang, Benjamin Recht, and Ludwig Schmidt. Evaluating machine accuracy on imagenet. In *International Conference on Machine Learning*, pages 8634–8644. PMLR, 2020.

[201] Jamie Shotton, Andrew Fitzgibbon, Mat Cook, Toby Sharp, Mark Finocchio, Richard Moore, Alex Kipman, and Andrew Blake. Real-time human pose recognition in parts from single depth images. In *CVPR 2011*, pages 1297–1304. IEEE, 2011.

[202] Tomas Simon, Hanbyul Joo, Iain Matthews, and Yaser Sheikh. Hand keypoint detection in single images using multiview bootstrapping. In *Proceedings Of The IEEE Conference on Computer Vision and Pattern Recognition*, pages 1145–1153, 2017.

[203] Karen Simonyan and Andrew Zisserman. Two-stream convolutional networks for action recognition in videos. *Advances in Neural Information Processing Systems*, 27, 2014.

[204] Karen Simonyan and Andrew Zisserman. Very deep convolutional networks for large-scale image recognition. *ArXiv Preprint ArXiv:1409.1556*, 2014.

[205] Jascha Sohl-Dickstein, Eric Weiss, Niru Maheswaranathan, and Surya Ganguli. Deep unsupervised learning using nonequilibrium thermodynamics. In *International Conference on Machine Learning*, pages 2256–2265. PMLR, 2015.

[206] Qingyu Song, Changan Wang, Zhengkai Jiang, Yabiao Wang, Ying Tai, Chengjie Wang, Jilin Li, Feiyue Huang, and Yang Wu. Rethinking counting and localization in crowds: A purely point-based framework. In *Proceedings of the IEEE/CVF International Conference on Computer Vision*, pages 3365–3374, 2021.

[207] Yang Song, Jascha Sohl-Dickstein, Diederik P Kingma, Abhishek Kumar, Stefano Ermon, and Ben Poole. Score-based generative modeling through stochastic differential equations. *ArXiv Preprint ArXiv:2011.13456*, 2020.

[208] Ke Sun, Bin Xiao, Dong Liu, and Jingdong Wang. Deep high-resolution representation learning for human pose estimation. In *Proceedings of the IEEE/CVF Conference on Computer Vision and Pattern Recognition*, pages 5693–5703, 2019.

[209] Lin Sun, Kui Jia, Kevin Chen, Dit-Yan Yeung, Bertram E Shi, and Silvio Savarese. Lattice long short-term memory for human action recognition. In *Proceedings of the IEEE International Conference on Computer Vision*, pages 2147–2156, 2017.

[210] Wen-Tsai Sung and Yao-Chi Hsu. Designing an industrial real-time measurement and monitoring system based on embedded system and zigbee. *Expert Systems with Applications*, 38(4):4522–4529, 2011.

[211] Christian Szegedy, Wojciech Zaremba, Ilya Sutskever, Joan Bruna, Dumitru Erhan, Ian Goodfellow, and Rob Fergus. Intriguing properties of neural networks. *ArXiv Preprint ArXiv:1312.6199*, 2013.

[212] Mingxing Tan, Ruoming Pang, and Quoc V Le. Efficientdet: Scalable and efficient object detection. In *Proceedings of the IEEE/CVF Conference on Computer Vision and Pattern Recognition*, pages 10781–10790, 2020.

[213] Simen Thys, Wiebe Van Ranst, and Toon Goedemé. Fooling automated surveillance cameras: adversarial patches to attack person detection. In *Proceedings of the IEEE/CVF Conference on Computer Vision and Pattern Recognition Workshops*, pages 0–0, 2019.

[214] Antonio Tilocca, Paolo Borzone, Stefano Carosio, and Antonio Durante. Detecting fabric defects with a neural network using two kinds of optical patterns. *Textile Research Journal*, 72(6):545–550, 2002.

[215] Alexander Toshev and Christian Szegedy. Deeppose: Human pose estimation via deep neural networks. In *Proceedings of the IEEE Conference on Computer Vision and Pattern Recognition*, pages 1653–1660, 2014.

[216] Hugo Touvron, Matthieu Cord, Matthijs Douze, Francisco Massa, Alexandre Sablayrolles, and Hervé Jégou. Training data-efficient image transformers & distillation through attention. In *International Conference on Machine Learning*, pages 10347–10357. PMLR, 2021.

[217] Hugo Touvron, Thibaut Lavril, Gautier Izacard, Xavier Martinet, Marie-Anne Lachaux, Timothée Lacroix, Baptiste Rozière, Naman Goyal, Eric Hambro, Faisal Azhar, Aurélien Rodriguez, Armand Joulin, Edouard Grave, and Guillaume Lample. Llama: Open and efficient foundation language models. *ArXiv*, abs/2302.13971, 2023.

[218] Du Tran, Lubomir Bourdev, Rob Fergus, Lorenzo Torresani, and Manohar Paluri. Learning spatiotemporal features with 3D convolutional networks. In *Proceedings of the IEEE International Conference on Computer Vision*, pages 4489–4497, 2015.

[219] Ashish Vaswani, Noam Shazeer, Niki Parmar, Jakob Uszkoreit, Llion Jones, Aidan N Gomez, Łukasz Kaiser, and Illia Polosukhin. Attention is all you need. *Advances in Neural Information Processing Systems*, 30, 2017.

[220] Paul Viola and Michael J Jones. Robust real-time face detection. *International Journal of Computer Vision*, 57(2):137–154, 2004.

[221] Ao Wang, Hui Chen, Lihao Liu, Kai Chen, Zijia Lin, Jungong Han, and Guiguang Ding. Yolov10: Real-time end-to-end object detection. *ArXiv Preprint ArXiv:2405.14458*, 2024.

[222] Boyu Wang, Huidong Liu, Dimitris Samaras, and Minh Hoai. Distribution matching for crowd counting. In *Advances in Neural Information Processing Systems*, pages 1595–1607, 2020.

[223] Chien-Yao Wang, Alexey Bochkovskiy, and Hong-Yuan Mark Liao. Scaled-yolov4: Scaling cross stage partial network. In *Proceedings of the IEEE/CVF Conference on Computer Vision and Pattern Recognition*, pages 13029–13038, 2021.

[224] Chien-Yao Wang, Hong-Yuan Mark Liao, Yueh-Hua Wu, Ping-Yang Chen, Jun-Wei Hsieh, and I-Hau Yeh. Cspnet: A new backbone that can enhance learning capability of cnn. In *Proceedings of the IEEE/CVF Conference on Computer Vision and Pattern Recognition Workshops*, pages 390–391, 2020.

[225] Chien-Yao Wang, Hong-Yuan Mark Liao, Yueh-Hua Wu, Ping-Yang Chen, Jun-Wei Hsieh, and I-Hau Yeh. Cspnet: A new backbone that can enhance learning capability of cnn. In *Proceedings of the IEEE/CVF Conference on Computer Vision and Pattern Recognition Workshops*, pages 390–391, 2020.

[226] Chuan Wang, Hua Zhang, Liang Yang, Si Liu, and Xiaochun Cao. Deep people counting in extremely dense crowds. In *Proceedings of the 23rd ACM International Conference on Multimedia*, pages 1299–1302, 2015.

[227] Jianguo Wang, Xiaomeng Yi, Rentong Guo, Hai Jin, Peng Xu, Shengjun Li, Xiangyu Wang, Xiangzhou Guo, Chengming Li, and Xiaohai Xu. Milvus: A purpose-built vector data management system. In *SIGMOD/PODS '21: International Conference on Management Of Data*, 2021.

[228] Jingdong Wang, Ke Sun, Tianheng Cheng, Borui Jiang, Chaorui Deng, Yang Zhao, Dong Liu, Yadong Mu, Mingkui Tan, Xinggang Wang, et al. Deep high-resolution representation learning for visual recognition. *IEEE Transactions on Pattern Analysis and Machine Intelligence*, 43(10):3349–3364, 2020.

[229] Kun Wang, Zixuan Teng, and Tengyue Zou. Metal defect detection based on yolov5. In *Journal of Physics: Conference Series*, volume 2218, page 012050. IOP Publishing, 2022.

[230] Qi Wang, Junyu Gao, Wei Lin, and Xuelong Li. Nwpu-crowd: A large-scale benchmark for crowd counting and localization. *IEEE Transactions on Pattern Analysis and Machine Intelligence*, 43(6):2141–2149, 2020.

[231] Wenhai Wang, Enze Xie, Xiang Li, Deng-Ping Fan, Kaitao Song, Ding Liang, Tong Lu, Ping Luo, and Ling Shao. Pyramid vision transformer: A versatile backbone for dense prediction without convolutions. In *Proceedings of the IEEE/CVF International Conference on Computer Vision*, pages 568–578, 2021.

[232] Shih-En Wei, Varun Ramakrishna, Takeo Kanade, and Yaser Sheikh. Convolutional pose machines. In *CVPR*, 2016.

[233] Mathias Wien. High efficiency video coding. *Coding Tools and Specification*, 24:1, 2015.

[234] Oliver Wulf and Bernardo Wagner. Fast 3D scanning methods for laser measurement systems. In *International Conference on Control Systems and Computer Science (CSCS14)*, pages 2–5, 2003.

[235] Qingzhen Xu, Guangyi Huang, Mengjing Yu, and Yanliang Guo. Fall prediction based on key points of human bones. *Physica A: Statistical Mechanics and Its Applications*, 540:123205, 2020.

[236] Rongge Xu, Ruiyang Hao, and Biqing Huang. Efficient surface defect detection using self-supervised learning strategy and segmentation network. *Advanced Engineering Informatics*, 52:101566, 2022.

[237] Apichet Yajai, Annupan Rodtook, Krisana Chinnasarn, and Suwanna Rasmequan. Fall detection using directional bounding box. In *2015 12th International Joint Conference on Computer Science and Software Engineering (JCSSE)*, pages 52–57. IEEE, 2015.

[238] Rikiya Yamashita, Mizuho Nishio, Richard Kinh Gian Do, and Kaori Togashi. Convolutional neural networks: an overview and application in radiology. *Insights into Imaging*, 9:611–629, 2018.

[239] Sijie Yan, Yuanjun Xiong, and Dahua Lin. Spatial temporal graph convolutional networks for skeleton-based action recognition. In *Proceedings of the AAAI Conference on Artificial Intelligence*, volume 32, 2018.

[240] Sen Yang, Zhibin Quan, Mu Nie, and Wankou Yang. Transpose: Keypoint localization via transformer. In *Proceedings of the IEEE/CVF International Conference on Computer Vision*, pages 11802–11812, 2021.

[241] Shuo Yang, Ping Luo, Chen-Change Loy, and Xiaoou Tang. Wider face: A face detection benchmark. In *Proceedings of the IEEE Conference on Computer Vision and Pattern Recognition*, pages 5525–5533, 2016.

[242] Hu Ye, Jun Zhang, Sibo Liu, Xiao Han, and Wei Yang. Ip-adapter: Text compatible image prompt adapter for text-to-image diffusion models. *ArXiv Preprint ArXiv:2308.06721*, 2023.

[243] Lemei Yin. A review of text-to-image synthesis methods. In *2024 5th International Conference on Computer Vision, Image and Deep Learning (CVIDL)*, pages 858–861. IEEE, 2024.

[244] Jiawei Yu, Ye Zheng, Xiang Wang, Wei Li, Yushuang Wu, Rui Zhao, and L Fast-Flow Wu. Unsupervised anomaly detection and localization via 2D normalizing flows. In *Proceedings of the Conference on Computer Vision and Pattern Recognition (CVPR), Nashville, TN, USA*, pages 19–25, 2021.

[245] Mingqi Yu, Hongwei Yue, Chuangquan Chen, Cong Zhou, Yufeng Huang, Dong Wang, and Fuqin Deng. An accurate defect detection model for photosensitive elements based on improved yolov4 deep neural network. In *2022 5th International Conference on Pattern Recognition and Artificial Intelligence (PRAI)*, pages 337–345. IEEE, 2022.

[246] Yifei Yuan and Xiaowu Zhao. 5g: Vision, scenarios and enabling technologies. *ZTE Communications*, 13(1):3, 2015.

[247] Daoguang Zan, Bei Chen, Fengji Zhang, Dianjie Lu, Bingchao Wu, Bei Guan, Wang Yongji, and Jian-Guang Lou. Large language models meet NL2Code: A survey. In *Proceedings of the 61st Annual Meeting of the Association for Computational Linguistics (Volume 1: Long Papers)*, pages 7443–7464, Toronto, Canada, July 2023. Association for Computational Linguistics.

[248] Chengliang Zhang, Tianhui Li, and Wenbin Zhang. The detection of impurity content in machine-picked seed cotton based on image processing and improved YOLO V4. *Agronomy*, 12(1):66, 2021.

[249] Chunhui Zhang, Li Liu, Yawen Cui, Guanjie Huang, Weilin Lin, Yiqian Yang, and Yuehong Hu. A comprehensive survey on segment anything model for vision and beyond. *ArXiv Preprint ArXiv:2305.08196*, 2023.

[250] Dongyan Zhang, Cheng Jiao, and Shuo Wang. Smoking image detection based on convolutional neural networks. In *2018 IEEE 4th International Conference on Computer and Communications (ICCC)*, pages 1509–1515. IEEE, 2018.

[251] Hongguang Zhang, Li Zhang, Xiaojuan Qi, Hongdong Li, Philip HS Torr, and Piotr Koniusz. Few-shot action recognition with permutation-invariant attention. In *Computer Vision–ECCV 2020: 16th European Conference, Glasgow, UK, August 23–28, 2020, Proceedings, Part V 16*, pages 525–542. Springer, 2020.

[252] Jin Zhang, Cheng Wu, and Yiming Wang. Human fall detection based on body posture spatio-temporal evolution. *Sensors*, 20(3):946, 2020.

[253] Rui Zhang, Zheng Zhu, Peng Li, Rui Wu, Chaoxu Guo, Guan Huang, and Hailun Xia. Exploiting offset-guided network for pose estimation and tracking. In *CVPR Workshops*, pages 20–28, 2019.

[254] Shanshan Zhang, Christian Bauckhage, and Armin B Cremers. Informed haar-like features improve pedestrian detection. In *Proceedings of the IEEE Conference on Computer Vision and Pattern Recognition*, pages 947–954, 2014.

[255] Shanshan Zhang, Rodrigo Benenson, and Bernt Schiele. Citypersons: A diverse dataset for pedestrian detection. In *Proceedings of the IEEE Conference on Computer Vision and Pattern Recognition*, pages 3213–3221, 2017.

[256] Shanshan Zhang, Rodrigo Benenson, Bernt Schiele, et al. Filtered channel features for pedestrian detection. In *Proceedings of the IEEE Conference on Computer Vision and Pattern Recognition*, pages 1751–1760, 2015.

[257] Shanshan Zhang, Lihong He, Eduard Dragut, and Slobodan Vucetic. How to invest my time: Lessons from human-in-the-loop entity extraction. In *Proceedings of the 25th ACM SIGKDD International Conference on Knowledge Discovery & Data Mining*, pages 2305–2313, 2019.

[258] Shifeng Zhang, Cheng Chi, Yongqiang Yao, Zhen Lei, and Stan Z Li. Bridging the gap between anchor-based and anchor-free detection via adaptive training sample selection. In *Proceedings of the IEEE/CVF Conference on Computer Vision and Pattern Recognition*, pages 9759–9768, 2020.

[259] Yifu Zhang, Peize Sun, Yi Jiang, Dongdong Yu, Fucheng Weng, Zehuan Yuan, Ping Luo, Wenyu Liu, and Xinggang Wang. Bytetrack: Multi-object tracking by associating every detection box. In *European Conference on Computer Vision*, pages 1–21. Springer, 2022.

[260] Yingying Zhang, Desen Zhou, Siqin Chen, Shenghua Gao, and Yi Ma. Single-image crowd counting via multi-column convolutional neural network. In *Proceedings of the IEEE Conference on Computer Vision and Pattern Recognition*, pages 589–597, 2016.

[261] Zijun Zhang. Improved adam optimizer for deep neural networks. In *2018 IEEE/ACM 26th International Symposium on Quality of Service (IWQoS)*, pages 1–2. IEEE, 2018.

[262] Xingyi Zhou, Dequan Wang, and Philipp Krähenbühl. Objects as points. In arXiv preprint 2, 2019.

[263] Zongwei Zhou, Md Mahfuzur Rahman Siddiquee, Nima Tajbakhsh, and Jianming Liang. UNet++: A nested U-Net architecture for medical image segmentation. In *Deep Learning in Medical Image Analysis and Multimodal Learning for Clinical Decision Support: 4th International Workshop, DLMIA 2018, and 8th International Workshop, ML-CDS 2018, Held in Conjunction with MICCAI 2018, Granada, Spain, September 20, 2018, Proceedings 4*, pages 3–11. Springer, 2018.

[264] Chenchen Zhu, Yihui He, and Marios Savvides. Feature selective anchor-free module for single-shot object detection. In *Proceedings of the IEEE/CVF Conference on Computer Vision and Pattern Recognition*, pages 840–849, 2019.

[265] Jun-Yan Zhu, Taesung Park, Phillip Isola, and Alexei A Efros. Unpaired image-to-image translation using cycle-consistent adversarial networks. In *Proceedings of the IEEE International Conference on Computer Vision*, pages 2223–2232, 2017.

[266] Zhuangdi Zhu, Kaixiang Lin, Anil K Jain, and Jiayu Zhou. Transfer learning in deep reinforcement learning: A survey. *IEEE Transactions on Pattern Analysis and Machine Intelligence*, 2023.

[267] Michael Zollhöfer, Patrick Stotko, Andreas Görlitz, Christian Theobalt, Matthias Nießner, Reinhard Klein, and Andreas Kolb. State of the art on 3D reconstruction with RGB-D cameras. In *Computer Graphics Forum*, volume 37, pages 625–652. Wiley Online Library, 2018.

# Index

For Product Safety Concerns and Information please contact our EU
representative  GPSR@taylorandfrancis.com
Taylor & Francis Verlag GmbH, Kaufingerstraße 24, 80331 München, Germany

www.ingramcontent.com/pod-product-compliance
Lightning Source LLC
Chambersburg PA
CBHW082007190326
41458CB00010B/3099